T0176674

Wavelet Neural Networks

Wavelet Neural Networks

With Applications in Financial Engineering, Chaos, and Classification

Antonios K. Alexandridis
School of Mathematics, Statistics and Actuarial Science
University of Kent
Canterbury, United Kingdom

Achilleas D. Zapranis
Department of Accounting and Finance
University of Macedonia
Thessaloniki, Greece

Published by John Wiley & Sons, Inc., Hoboken, New Jersey.
Published simultaneously in Canada.

For general information on our other products and services or for technical support, please contact our Customer Care Department within the United States at (800) 762-2974, outside the United States at (317) 572-3993 or fax (317) 572-4002.

Wiley also publishes its books in a variety of electronic formats. Some content that appears in print may not be available in electronic formats. For more information about Wiley products, visit our web site at www.wiley.com.

Library of Congress Cataloging-in-Publication Data:

Alexandridis, Antonios K.
  Wavelet neural networks : with applications in financial engineering, chaos, and classification / Antonios K. Alexandridis, School of Mathematics, Statistics and Actuarial Science, University of Kent, Canterbury, UK, Achilleas D. Zapranis, Department of Accounting and Finance, University of Macedonia, Thessaloniki, Greece.
      pages cm
  Includes bibliographical references and index.
  ISBN 978-1-118-59252-6 (cloth)
  1. Wavelets (Mathematics)   2. Neural networks (Computer science)
3. Financial engineering.   I. Zapranis, Achilleas, 1965–   II. Title.
  QA403.3.A428 2014
  006.3'2–dc23
                                            2013047838

Printed in the United States of America

10 9 8 7 6 5 4 3 2 1

To our families

Contents

Preface

Wavelet networks are a new class of networks that combine classic sigmoid neural networks and wavelet analysis. Wavelet networks were proposed as an alternative to feedforward neural networks, which would alleviate the weaknesses associated with wavelet analysis and neural networks while preserving the advantages of each method.

Recently, wavelet networks have gained a lot of attention and have been used with great success in a wide range of applications: financial modeling; engineering; system control; short-term load forecasting; time-series prediction; signal classification and compression; signal denoising; static, dynamic, and nonlinear modeling; and nonlinear static function approximation—to mention some of the most important.

However, a major weakness of wavelet neural modeling is the lack of a generally accepted framework for applying wavelet networks. The purpose of this book is to present a step-by-step guide for model identification for wavelet networks. We describe a complete statistical model identification framework for applying wavelet networks in a variety of ways. Although vast literature on wavelet networks exists, to our knowledge this is the first study that presents a step-by-step guide for model identification for wavelet networks. Model identification can be separated into two parts: model selection and variable significance testing.

A concise and rigorous treatment for constructing optimal wavelet networks is provided. More precisely, the following subjects are examined thoroughly: the structure of a wavelet network; training methods; initialization algorithms; variable significance and variable selection algorithms; model selection methods; and methods to construct confidence and prediction intervals. The book links the mathematical aspects of the construction of wavelet network to modeling and forecasting applications in finance, chaos, and classification. Wavelet networks can constitute a valuable tool in financial engineering since they make no a priori assumptions about the nature of the dynamics that govern financial time series. Although we employ wavelet networks primarily in financial applications, it is clear that they can be utilized in modeling any nonlinear function. Hence, researchers can apply wavelet networks in any discipline to model any nonlinear problem.

Our goal has been to make the material accessible and readable without excessive mathematical requirements: for example, at the level of advanced M.B.A. or Ph.D. students. There is an introduction or tutorial to acquaint nonstatisticians with the basic principles of wavelet analysis, and a similar but more extensive introduction to neural networks for noncomputer scientists: first introducing them as regression models and gradually building up to more complex frameworks.

Familiarity with wavelet analysis, neural wavelets, or wavelet networks will help, but it is not a prerequisite. The book will take the reader to the level where he or she is expected to be able to utilize the proposed methodologies in applying wavelet networks to model various applications.

The book is meant to be used by a wide range of practitioners:

- By quantitative and technical analysts in investment institutions such as banks, insurance companies, securities houses, companies with intensive international activities, and financial consultancy firms, as well as fund managers and institutional investors.

- By those in such fields as engineering, chemistry, and biomedicine.

- By students in advanced postgraduate programs in finance, M.B.A., and mathematical modeling courses, as well as in computational economics, informatics, decision science, finance, artificial intelligence, and computational finance. It is anticipated that a considerable segment of the readership will originate from within the neural network application community as well as from students in the mathematical, physical, and engineering sciences seeking employment in the mathematical modeling services.

- By researchers in identification and modeling for complex nonlinear systems, wavelet neural networks, artificial intelligence, mathematical modeling, and relevant Ph.D. programs.

Supplementary material for this book may be found by entering ISBN 9781118592526 at booksupport.wiley.com.

During the preparation of the book, the help of my (A.K.A.) wife, Christina Ioannidou, was significant, and we would like to thank her for her careful reading of the manuscript.

Canterbury, UK   ANTONIOS K. ALEXANDRIDIS
Thessaloniki, Greece   ACHILLEAS D. ZAPRANIS
October 2013

1

Machine Learning and Financial Engineering

Wavelet networks are a new class of networks that combine the classic sigmoid neural networks and wavelet analysis. Wavelet networks were proposed by Zhang and Benveniste (1992) as an alternative to feedforward neural networks which would alleviate the weaknesses associated with wavelet analysis and neural networks while preserving the advantages of each method.

Recently, wavelet networks have gained a lot of attention and have been used with great success in a wide range of applications, ranging from engineering; control; financial modeling; short-term load forecasting; time-series prediction; signal classification and compression; signal denoising; static, dynamic, and nonlinear modeling; to nonlinear static function approximation.

Wavelet networks are a generalization of radial basis function networks (RBFNs). Wavelet networks are hidden layer networks that use a wavelet for activation instead of the classic sigmoidal family. It is important to mention here that multidimensional wavelets preserve the "universal approximation" property that characterizes neural networks. The nodes (or *wavelons*) of wavelet networks are wavelet coefficients of the function expansion that have a significant value. In Bernard et al. (1998), various reasons were presented for why wavelets should be used instead of other transfer functions. In particular, first, wavelets have high compression abilities, and second, computing the value at a single point or updating a function estimate from a new local measure involves only a small subset of coefficients.

Wavelet Neural Networks: With Applications in Financial Engineering, Chaos, and Classification,
First Edition. Antonios K. Alexandridis and Achilleas D. Zapranis.
© 2014 John Wiley & Sons, Inc. Published 2014 by John Wiley & Sons, Inc.

In statistical terms, wavelet networks are nonlinear nonparametric estimators. Moreover, the universal approximation property states that wavelet networks can approximate, to any degree of accuracy, any nonlinear function and its derivatives. The useful properties of wavelet networks make them an excellent nonlinear estimator for modeling, interpreting, and forecasting complex financial problems and phenomena when only speculation is available regarding the underlying mechanism that generates possible observations.

In the context of a globalized economy, companies that offer financial services try to establish and maintain their competitiveness. To do so, they develop and apply advanced quantitative methodologies. Neural networks represent a new and exciting technology with a wide range of potential financial applications, ranging from simple tasks of assessing credit risk to strategic portfolio management. The fact that neural and wavelet networks avoid a priori assumptions about the evolution in time of the various financial variables makes them a valuable tool.

The purpose of this book is to present a step-by-step guide for model identification of wavelet networks. A generally accepted framework for applying wavelet networks is missing from the literature. In this book we present a complete statistical model identification framework to utilize wavelet networks in various applications. More precisely, wavelet networks are utilized for time-series prediction, construction of confidence and prediction intervals, classification and modeling, and forecasting of chaotic time series in the context of financial engineering. Although our proposed framework is examined primarily for its use in financial applications, it is not limited to finance. It is clear that it can be adopted and used in any discipline in the context of modeling any nonlinear problem or function.

The basic introductory notions are presented below. Fist, financial engineering and its relationship to machine learning and wavelet networks are discussed. Next, research areas related to financial engineering and its function and applications are presented. The basic notions of wavelet analysis and of neural and wavelet networks are also presented. More precisely, the basic mathematical notions that will be needed in later chapters are presented briefly. Also, applications of wavelet networks in finance are presented. Finally, the basic aspects of the framework proposed for the construction of optimal wavelet networks are discussed. More precisely, model selection, variable selection, and model adequacy testing stages are introduced.

FINANCIAL ENGINEERING

The most comprehensive definition of financial engineering is the following: *Financial engineering* involves the design, development, and implementation of innovative financial instruments and processes, and the formulation of creative solutions to problems of finance (Finnerty, 1988). From the definition it is clear that financial engineering is linked to innovation. A general definition of financial innovation includes not only the creation of new types of financial instruments, but the development and evolution of new financial institutions (Mason et al., 1995). Financial innovation is the driving force behind the financial system in fulfilling its primary

FINANCIAL ENGINEERING AND RELATED RESEARCH AREAS

function: the most efficient possible allocation of financial resources (Ζαπράνης, 2005). Investors, organizations, and companies in the financial sector benefit from financial innovation. These benefits are reflected in lower funding costs, improved yields, better management of various risks, and effective operation within changing regulations.

In recent decades the use of mathematical techniques and processes, derived from operational research, has increased significantly. These methods are used in various aspects of financial engineering. Methods such as decision analysis, statistical estimation, simulation, stochastic processes, optimization, decision support systems, neural networks, wavelet networks, and machine learning in general have become indispensable in several domains of financial operations (Mulvey et al., 1997).

According to Marshall and Bansal (1992), many factors have contributed to the development of financial engineering, including technological advances, globalization of financial markets, increased competition, changing regulations, the increasing ability to solve complex financial models, and the increased volatility of financial markets. For example, the operation of the derivatives markets and risk management systems is supported decisively by continuous advances in the theory of the valuation of derivatives and their use in hedging financial risks. In addition, the continuous increase in computational power while its cost is being reduced makes it possible to monitor thousands of market positions in real time to take advantage of short-term anomalies in the market.

In addition to their knowledge of economic and financial theory, financial engineers are required to possess the quantitative and technical skills necessary to implement engineering methods to solve financial problems. Financial engineering is a unique field of finance that does not necessarily focus on people with advanced technical backgrounds who wish to move into the financial area but, is addressed to those who wish to get involved in investment banking, investment management, or risk management.

There is a mistaken point of view that financial engineering is accessible only by people who have a strong mathematical and technical background. The usefulness of a financial innovation should be measured on the basis of its effect on the efficiency of the financial system, not on the degree of novelty that introduces. Similarly, the power of financial engineering should not be considered in the light of the complexity of the models that are used but from the additional administrative and financial flexibility that it offers its users. Hence, financial engineering is addressed to a large audience and should be considered within the broader context of the administrative decision-making system that it supports.

FINANCIAL ENGINEERING AND RELATED RESEARCH AREAS

Financial engineering is a very large multidisciplinary field of research. As a result, researchers are often focused on smaller subfields of financial engineering. There are two main branches of financial engineering: *quantitative finance* and *financial econometrics*. Quantitative finance is a combination of two very important and

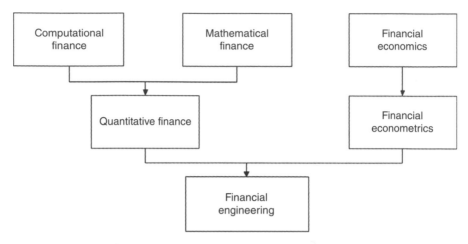

Figure 1.1   *Research areas related to financial engineering.*

popular subfields of finance: *mathematical finance* and *computational finance*. On the other hand, financial econometrics arises from *financial economics*. Research areas related to financial engineering are illustrated in Figure 1.1.

The scientific field of financial engineering is closely related to the relevant disciplinary areas of mathematical finance and computational finance, as all focus on the use of mathematics, algorithms, and computers to solve financial problems. It can be said that financial engineering is a multidisciplinary field involving financial theory, the methods of engineering, the tools of mathematics, and the practice of programming. However, financial engineering is focused on applications, whereas mathematical finance has a more theoretical perspective.

Mathematical finance, a field of applied mathematics concerned with financial markets, began in the 1970s. Its primary focus was the study of mathematics applied to financial concerns. Today, mathematical finance is an established and very important autonomous field of knowledge. In general, financial mathematicians study a problem and try to derive a mathematical or numerical model by observing the output values: for example, market prices. Their analysis does not necessarily have a link back to financial theory. More precisely, mathematical consistency is required, but not necessarily compatibility with economic theory.

Mathematical finance is closely related to computational finance. More precisely, the two fields overlap. Mathematical finance deals with the development of financial models, and computational finance is concerned with their application in practice. Computational finance emphasizes practical numerical methods rather than mathematical proofs, and focuses on techniques that apply directly to economic analyses. In addition to a good knowledge of financial theory, the background of people working in the field of computational finance combines fluency in fields such as algorithms, networks, databases, and programming languages (e.g., C/C++, Java, Fortran).

Today, the disciplinary area of mathematical finance and computational finance constitutes part of a larger, established, and more general area of finance called *quantitative finance*. In general, there are two main areas in which advanced mathematical and computational techniques are used in finance. One tries to derive mathematical formulas for the prices of derivatives, the other one deals with risk and portfolio management.

Financial econometrics is another field of knowledge closely related (although more remote) to financial engineering. Financial econometrics is the basic method of inference in the branch of economics termed financial economics. More precisely, the focus is on decisions made under uncertainty in the context of portfolio management and their implications to the valuation of securities (Huang and Litzenberger, 1988). The objective is to analyze financial models empirically under the assumption of uncertainty in the decisions of investors and hence in market prices. For example, the martingale model for capital asset pricing is related to mathematical finance. However, the empirical analysis of the behavior of the autocorrelation coefficient of the price changes generated by the martingale model is the subject of financial econometrics.

We illustrate the various subfields of financial engineering by the following example. A financial economist studies the structural reasons that a company may have a certain share price. A financial mathematician, on the other hand, takes the share price as a given and may use a stochastic model in an attempt to derive the corresponding price of a derivative with the stock as an underlying asset. The fundamental theorem of arbitrage-free pricing is one of the key theorems in mathematical finance, while the differential Black–Scholes–Merton approach (Black and Scholes, 1973) finds applications in the context of pricing options. However, to apply the stochastic model, a computational translation of the mathematics to a computing and numerical environment is necessary.

FUNCTIONS OF FINANCIAL ENGINEERING

Financial engineers are involved in many important functions in a financial institution. According to Mulvey et al. (1997), financial engineering is used widely in four major functions in finance: (1) corporate finance, (2) trading, (3) investment management, and (4) risk management (Figure 1.2). In corporate finance, large-scale businesses are interested in raising funds for their operation. Financial engineers develop new instruments or enhance existing ones in order to secure these funds. Also, they are involved in takeovers and buyouts. In trading of securities or derivatives, the objective of a financial engineer is to develop new dynamic trading strategies. In investment management the aim is to develop new investment vehicles for investors.

Examples presented by Mulvey et al. (1997) include high-yield mutual funds, money market funds, and the repo market. In addition, they develop systems for transforming high-risk investment instruments to low-risk instruments by applying techniques such as repackaging and overcollaterization. Finally, in risk management, a financial engineer must, on the one hand, assess the various types of risk of a range of securities and, on the other hand, use the appropriate methodologies and tools to construct portfolios with the desired levels of risk and return. These methodological approaches relate primarily to portfolio insurance, portfolio immunization

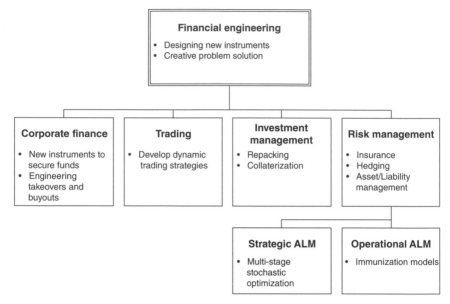

Figure 1.2   *Financial engineering activities according to Mulvey et al. (1997).*

against changes in certain financial variables, hedging, and efficient assets/liability management.

Risk management is a crucial part of corporate financial management. The interrelated areas of risk management and financial engineering find direct applications in many problems of corporate financial management, such as assessment of default risk, credit risk, portfolio selection and management, sovereign and country risk, and financial programing, to name a few.

During the past three decades, a series of new scientific tools derived from the wider field of operations research and artificial intelligence has been developed for the most realistic and comprehensive management of financial risks. Techniques that have been proposed and implemented include multicriteria decision analysis, expert systems, neural networks, genetic and evolutionary algorithms, fuzzy networks, and wavelet networks. A typical example is the use of neural networks by Zapranis and Sivridis (2003) to estimate the speed of inversion within the Vasicek model, used to derive the term structure of short-term interest rates.

APPLICATIONS OF MACHINE LEARNING IN FINANCE

Neural networks and machine learning in general are employed with considerable success in primarily three types of applications in finance: (1) modeling for classification and prediction, (2) associative memory, and (3) clustering (Hawley et al., 1990). The use of wavelet networks is shown in Figure 1.3.

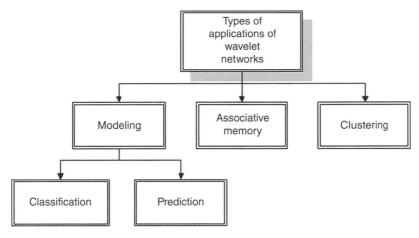

Figure 1.3   *Types of applications of wavelet networks.*

Classification includes assignment of units in predefined groups or classes, based on the discovery of regularities that exist in the sample. Generally, nonlinear nonparametric predictors such as wavelet networks are able to classify the units correctly even if the sample is incomplete or additional noise has been added. Typical examples of such applications are the visual recognition of handwritten characters and the identification of underwater targets by sonar. In finance, an example of a classification application could be the grouping of bonds, based on regularities in financial data of the issuer, into categories corresponding to the rating assigned by a specialized company. Other examples are the approval of credit granting (the decision as to who receives credit and how much), stock selection (classification based on the anticipated yield), and automated trading systems.

The term *prediction* refers to the development of mathematical relationships between the input variables of the wavelet network and usually one (although it can be more) output variable. Artificial networks expand the common techniques that are used in finance, such as linear and polynomial regression and autoregressive moving averages (ARMA and ARIMA). In finance, machine learning is used mainly in classification and prediction applications. When a wavelet network is trained, it can be used for the prediction of a financial time series. For example, a wavelet network can be used to produce point estimates of the future prices or returns of a particular stock or index. However, financial analysts are usually also interested in confidence and prediction intervals. For example, if the price of a stock moves outside the prediction interval, a financial analyst can adjust the trading strategy.

In *associative memory* applications the goal is to produce an output corresponding to the class or group desired, based on one input vector presented in the neural network that determines which output is to be produced. For example, the input vector may be a digitized image of a fingerprint, and the output desired may be reconstruction of the entire fingerprint.

Clustering is used to group a large number of different input variables, each of which, however, has some similarities with other inputs. Clustering is useful for compression or filtering of data without the loss of a substantial part of the information. A financial application could be the creation of clusters of corporate bonds that correspond to uniform risk classes based on data from financial statements. The number and composition of the classes will be determined by the model, not by the user. In this case, in contrast to the use of classification, the categories are not predetermined. This method could provide an investor with a diversified portfolio.

FROM NEURAL TO WAVELET NETWORKS

In this section the basic notions of wavelet analysis, neural networks, and wavelet networks are presented. Our purpose is to present the basic mathematical and theoretical background that is used in subsequent chapters. Also, the reasons that motivated the combination of wavelet analysis and neural networks to create a new tool, wavelet networks, are discussed.

Wavelet Analysis

Wavelet analysis is a mathematical tool used in various areas of research. Recently, wavelets have been used especially to analyze time series, data, and images. Time series are represented by local information such as frequency, duration, intensity, and time position, and by global information such as the mean states over different time periods. Both global and local information is needed for the correct analysis of a signal. The wavelet transform (WT) is a generalization of the Fourier transform (FT) and the windowed Fourier transform (WFT).

Fourier Transform   The attempt to understand complicated time series by breaking them into basic pieces that are easier to understand is one of the central themes in Fourier analysis. In the framework of Fourier series, complicated periodic functions are written as the sum of simple waves represented mathematically by sines and cosines. More precisely, Fourier transform breaks a signal down into a linear combination of constituent sinusoids of various frequencies; hence, the Fourier transform is decomposition on a frequency-by-frequency basis.

Let $f: \mathbb{R} \to \mathbb{C}$ be a periodic function with period $T > 0$ that satisfies

$$\|f\|^2 = \int_{-\infty}^{\infty} |f|^2 \, dt < \infty \tag{1.1}$$

Then its FT is given by

$$\hat{f}(\omega) = \int_{-\infty}^{\infty} e^{-2\pi i \omega t} f(t) \, dt \tag{1.2}$$

and its Fourier coefficients are given by

$$c_n = \int_{-\infty}^{\infty} e^{-2\pi i \omega_n t} f(t)\, dt \tag{1.3}$$

where $\omega_n = n/T$ and

$$e^{2\pi i \omega_n} = \cos 2\pi \omega_n + i \sin 2\pi \omega_n \tag{1.4}$$

In a common interpretation of the FT given by Mallat (1999), the periodic function $f(t)$ is considered as a musical tone that the FT decomposes to a linear combination of different notes c_n with frequencies ω_n. This method allows us to compress the original signal, in the sense that it is not necessary to store the entire signal; only the coefficients and the corresponding frequencies are required. Knowing the coefficients c_n, one can synthesize the original signal $f(t)$. This procedure, called *reconstruction*, is achieved by the inverse FT, given by

$$f(t) = \int_{-\infty}^{\infty} e^{2\pi i \omega t} \hat{f}(\omega)\, d\omega \tag{1.5}$$

The FT has been used successfully in a variety of applications. The most common use of FT is in solving partial differential equations (Bracewell, 2000), in image processing and filtering (Lim, 1990), in data processing and analysis (Oppenheim et al., 1999), and in optics (Wilson, 1995).

Short-Time Fourier Transform (Windowed Fourier) Fourier analysis performs extremely well in the analysis of periodic signals. However, in transforming to the frequency domain, time information is lost. When looking at the Fourier transform of a signal, it is impossible to tell when a particular event took place. This is a serious drawback if the signal properties change a lot over time: that is, if they contain nonstationary or transitory characteristics: drift, trends, abrupt changes, or beginnings and ends of events. These characteristics are often the most important part of a time series, and Fourier transform is not suited to detecting them (Zapranis and Alexandridis, 2006).

Trying to overcome the problems of classical Fourier transform, Gabor applied the Fourier transform in small time "windows" (Mallat, 1999). To achieve a sort of compromise between frequency and time, Fourier transform was expanded in windowed Fourier transform or short-time Fourier transform (STFT). WFT uses a window across the time series and then uses the FT of the windowed series. This is a decomposition of two parameters, time and frequency. Window Fourier transform is an extension of the Fourier transform where a symmetric window, $g(u) = g(-u)$, is used to localize signals in time. If $t \in \mathbb{R}$, we define

$$f_t(u) = \bar{g}(u - t) f(u) \tag{1.6}$$

Expression (1.6) reveals that $f_t(u)$ is a localized version of f that depends only on values of $f(u)$. Again following the notation of Kaiser (1994), the STFT of f is given by

$$\tilde{f}(\omega, t) = \hat{f}_t(\omega) = \int_{-\infty}^{\infty} e^{-2\pi i \omega u} \, \bar{g}(u - t) f(u) \, du \qquad (1.7)$$

It is easy to see that by setting $g(u) = 1$, the SFTF is reduced to ordinary FT. Because of the similarity of equations (1.2) and (1.7), the inverse SFTF can be defined as

$$f(u) = C^{-1} \int_{-\infty}^{\infty} \int_{-\infty}^{\infty} e^{2\pi i \omega u} g\,(u - t)\tilde{f}\,(\omega, t) \, d\omega \, dt \qquad (1.8)$$

where $C = \|g\|^2$.

As mentioned earlier, FT can be used to analyze a periodic musical tone. However, if the musical tone is not periodic but rather is a series of notes or a melody, the Fourier series cannot be used directly (Kaiser, 1994). On the other hand, the STFT can analyze the melody and decompose it to notes, but it can also give the information when a given note ends and the next one begins. The STFT has been used successfully in a variety of applications. Common uses are in speech processing and spectral analysis (Allen, 1982) and in acoustics (Nawab et al., 1983), among others.

Extending the Fourier Transform: The Wavelet Analysis Paradigm

As mentioned earlier, Fourier analysis is inefficient in dealing with the local behavior of signals. On the other hand, windowed Fourier analysis is an inaccurate and inefficient tool for analyzing regular time behavior that is either very rapid or very slow relative to the size of the window (Kaiser, 1994). More precisely, since the window size is fixed with respect to frequency, WFT cannot capture events that appear outside the width of the window. Many signals require a more flexible approach: that is, one where we can vary the window size to determine more accurately either time or frequency.

Instead of the constant window used in WFT, waveforms of shorter duration at higher frequencies and waveforms of longer duration at lower frequencies were used as windows by Grossmann and Morlet (1984). This method, called *wavelet analysis*, is an extension of the FT. The fundamental idea behind wavelets is to analyze according to scale. Low scale represents high frequency, while high scales represent low frequency. The wavelet transform (WT) not only is localized in both time and frequency but also overcomes the fixed time–frequency partitioning. The new time–frequency partition is long in time at low frequencies and long in frequency at high frequencies. This means that the WT has good frequency resolution for low-frequency events and good time resolution for high-frequency events. Also, the WT adapts itself to capture features across a wide range of frequencies. Hence, the WT can be used to analyze time series that contain nonstationary dynamics at many different frequencies (Daubechies, 1992).

In finance, wavelet analysis is considered a new powerful tool for the analysis of financial time series, and it is applied in a wide range of financial problems. One example is the daily returns time series, which is represented by local information such as frequency, duration, intensity, and time position, and by global information such as the mean states over different time periods. Both global and local information is needed for a correct analysis of the daily return time series. Wavelets have the ability to decompose a signal or a time series on different levels. As a result, this decomposition brings out the structure of the underlying signal as well as trends, periodicities, singularities, or jumps that cannot be observed originally.

Wavelet analysis decomposes a general function or signal into a series of (orthogonal) basis functions called *wavelets*, which have different frequency and time locations. More precisely, wavelet analysis decomposes time series and images into component waves of varying durations called wavelets, which are localized variations of a signal (Walker, 2008). As illustrated by Donoho and Johnstone (1994), the wavelet approach is very flexible in handling very irregular data series. Ramsey (1999) also comments that wavelet analysis has the ability to represent highly complex structures without knowing the underlying functional form, which is of great benefit in economic and financial research. A particular feature of the signal analyzed can be identified with the positions of the wavelets into which it is decomposed.

Recently, an increasing number of studies apply wavelet analysis to analyze financial time series. Wavelet analysis was used by Alexandridis and Hasan (2013) to estimate the systematic risk of CAPM using wavelet analysis to examine the meteor shower effects of the global financial crisis. Similarly, one recent research strand of CAPM has built an empirical modeling strategy centering on the issue of the multiscale nature of the systematic risk using a framework of wavelet analysis (Fernandez, 2006; Gençay et al., 2003, 2005; Masih et al., 2010, Norsworthy et al., 2000; Rabeh and Mohamed, 2011). Wavelet analysis has also been used to construct a modeling and pricing framework in the context of financial weather derivatives (Alexandridis and Zapranis 2013a,b; Zapranis and Alexandridis, 2008, 2009).

Moreover, wavelet analysis was used by In and Kim (2006a,b) to estimate the hedge ratio, and it was used by Fernandez (2005), and In and Kim (2007) to estimate the international CAPM. Maharaj et al. (2011) made a comparison of developed and emerging equity market return volatility at different time scales. The relationship between changes in stock prices and bond yields in the G7 countries was studied by Kim and In (2007), while Kim and In (2005) examined the relationship between stock returns and inflation using wavelet analysis. He et al. (2012) studied the value-at-risk in metal markets, while a wavelet-based assessment of the risk in emerging markets was presented by Rua and Nunes (2012). Finally, a wavelet-based method for modeling and predicting oil prices was presented by Alexandridis and Livanis (2008), Alexandridis et al. (2008), and Yousefi et al. (2005). Finally, a survey of the contribution of wavelet analysis in finance was presented by Ramsey (1999).

Wavelets A wavelet ψ is a waveform of effectively limited duration that has an average value of zero. The WA procedure adopts a particular wavelet function called a *mother wavelet*. A *wavelet family* is a set of orthogonal basis functions generated by dilation and translation of a compactly supported *scaling function* φ (or *father wavelet*), and a *wavelet function* ψ (or mother wavelet).

The father wavelets φ and mother wavelets ψ satisfy

$$\int \varphi(t)\,dt = 1 \tag{1.9}$$

$$\int \psi(t)\,dt = 0 \tag{1.10}$$

The wavelet family consists of *wavelet children* which are dilated and translated forms of a mother wavelet:

$$\psi_{a,b}(t) = \frac{1}{\sqrt{a}}\psi\left(\frac{t-b}{a}\right) \tag{1.11}$$

where *a* is the *scale* or *dilation parameter* and *b* is the *shift* or *translation parameter*.

The value of the scale parameter determines the level of stretch or compression of the wavelet. The term $1/\sqrt{a}$ normalizes $\|\psi_{a,b}(t)\| = 1$. In most cases we limit our choice of *a* and *b* values by using a discrete set, because calculating wavelet coefficients at every possible scale is computationally intensive. Temporal analysis is performed with a contracted high-frequency version of the mother wavelet, while frequency analysis is performed with a dilated, low-frequency version of the same mother wavelet. In other words, whereas Fourier analysis consists of breaking a signal up into sine waves of various frequencies, wavelet analysis is the breakup of a signal into shifted and scaled versions of the original (or mother) wavelet (Misiti et al., 2009). Wavelet decomposition is illustrated in Figure 1.4.

Two versions of the WT can be distinguished: the continuous wavelet transform (CWT) and the discrete wavelet transform (DWT). The difference between them lies in the set of scales and positions at which each transform operates. The CWT can operate at every scale. However, an upper bound is determined since CWT is extremely computationally expensive. Also, in CWT the analyzing wavelet is shifted

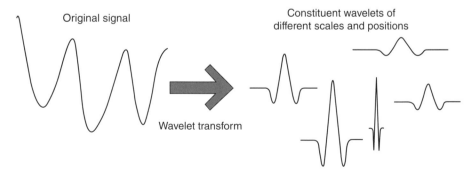

Figure 1.4   *Wavelet decomposition.*

smoothly over the full domain of the function analyzed. To reduce the computational burden, wavelet coefficients are calculated only on a subset of scales. This method is called the DWT.

In general, wavelets can be separated in orthogonal and nonorthogonal wavelets. The term *wavelet function* is used generically to refer to either orthogonal or nonorthogonal wavelets. An orthogonal set of wavelets is called a *wavelet basis*, and a set of nonorthogonal wavelets is termed a *wavelet frame*. The use of an orthogonal basis implies the use of the DWT, whereas frames can be used with either the discrete or the continuous transform.

Over the years a substantial number of wavelet functions have been proposed in the literature. The Gaussian, the Morlet, and the Mexican hat wavelets are crude wavelets that can be used only in continuous decomposition. The wavelets in the Meyer wavelet family are infinitely regular wavelets that can be used in both CWT and DWT. The wavelets in the Daubechies, symlet, and coiflet families are orthogonal and compactly supported wavelets. These wavelet families can also be used in CWT and DWT. The B-splines and biorthogonal wavelet families are biorthogonal, compactly supported wavelet pairs that can also be used in both CWT and DWT. Finally, the complex Gaussian, complex Morlet, complex Shannon, and complex-frequency B-spline wavelet families are complex wavelets that can be used in the complex CWT.

Generally, the DWT is used for data compression if the signal is already sampled, and the CWT is used for signal analysis. In the next sections the CWT and the DWT are examined in detail. In Figure 1.5 the Mexican hat wavelet is presented,

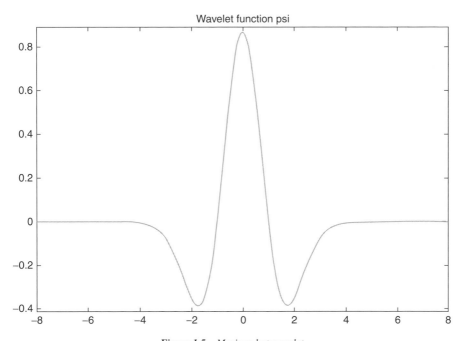

Figure 1.5 *Mexican hat wavelet.*

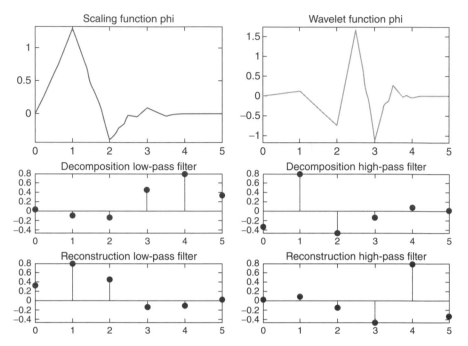

Figure 1.6 *Daubechies 3 wavelet (top right) and the scaling function (top left). The decomposition (middle) and reconstruction (bottom) filters are also presented with low-pass (left) and high-pass (right) filters.*

and in Figures 1.6 and 1.7 two wavelets of the Daubechies family are presented, the Daubechies 3 and 12, respectively. In addition, in Figures 1.6 and 1.7, the scaling functions and decomposition and reconstruction filters are presented with low- and high-pass filters.

Continuous Wavelet Transform Representing a signal as a function $f(t)$, the CWT of this function comprises the *wavelet coefficients, $C(a, b)$*, which are produced through the convolution of a mother wavelet function, $\psi(t)$, with the signal analyzed, $f(t)$. The CWT is defined as the summation over all time of the signal multiplied by scaled, shifted versions of the wavelet function:

$$\tilde{f}(a, b) = C(a, b) = \int_{-\infty}^{+\infty} \bar{\psi}_{a,b}(t) f(t)\, dt \tag{1.12}$$

where

$$\bar{\psi}_a(t) = \frac{1}{\sqrt{a}} \psi\left(\frac{-t}{a}\right) \tag{1.13}$$

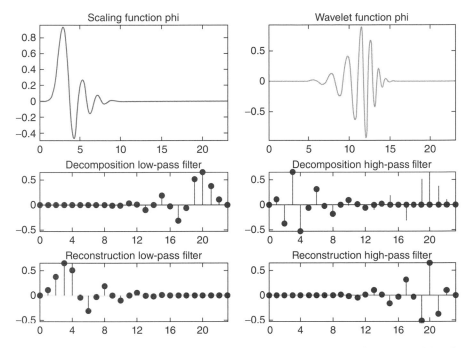

Figure 1.7 *The Daubechies 12 wavelet (top right) together with the scaling function (top left). The decomposition (middle) and reconstruction (bottom) filters are also presented with low-pass (left) and high-pass (right) filters.*

The CWT is continuous in the set of scales and positions at which it operates. It is also continuous in terms of shifting; during computation, the analyzing wavelet is shifted smoothly over the full domain of the function analyzed.

Either real or complex analytical wavelets can be used. Complex analytical wavelets can separate amplitude and phase components, while real wavelets are often used to detect sharp signal transitions. The results of the CWT are many wavelet coefficients C, which are a function of scale and position. Multiplying each coefficient by the appropriately scaled and shifted wavelet yields the constituent wavelets of the original signal. As the original signal can be represented in terms of a wavelet expansion (using coefficients in a linear combination of the wavelet functions), data operations can be performed using just the corresponding wavelet coefficients.

Similar to the STFT, the magnitude of wavelet coefficients will be represented by a plot on which the x-axis represents the position along time and the y-axis represents the scale. This plot, called a *scalogram*, represents the energy density of the signal:

$$W\tilde{f}(\omega, t) = |\tilde{f}(a, b)|^2 = \left| \int_{-\infty}^{+\infty} \bar{\psi}_{a,b}(t) f(t)\, dt \right|^2 \tag{1.14}$$

A scalogram allows the changing spectral composition of nonstationary signals to be measured and compared. If the wavelet $\psi \in \mathbf{L}^2\,(\mathbb{R})$ and satisfies the admissibility condition

$$C_\psi = \int_{-\infty}^{+\infty} \frac{|\hat{\psi}\,(\omega)|^2}{\omega} d\omega < +\infty \tag{1.15}$$

the original signal can be synthesized from the WT (Daubechies, 1992). The continuous reconstruction formula is given by

$$f(t) = \frac{1}{C_\psi} \int_{-\infty}^{+\infty} \int_{-\infty}^{+\infty} \frac{1}{a^2} \psi_{a,b}(t)\, da\, db \tag{1.16}$$

Discrete Wavelet Transform   Calculating wavelet coefficients at every possible scale is computationally expensive. However, if we choose only a subset of scales and translations based on powers of 2 (the *dyadic* lattice), our analysis will be much more efficient and just as accurate (Misiti et al., 2009). We obtain such an analysis from the DWT. In the DWT the wavelet family is taken from a double-indexed regular lattice:

$$\{(a_j, b_k) = (p^j, kqp^j)\colon j, k \in \mathbf{Z}\}$$

where the parameters p and q denote the step sizes of the dilation and the translation parameters. For $p = 2$ and $q = 1$ we have the standard dyadic lattice:

$$\{(a_j, b_k) = (2^j, 2^j k)\colon j, k \in \mathbf{Z}\} \tag{1.17}$$

Thus, the scaling function ϕ generates for each $j \in \mathbf{Z}$ the sets $V_j = \text{span}\{\phi_{j,k}, k \in \mathbb{Z}\}$, where $\mathbb{Z}$ denotes the set of integers and

$$\phi_{j,k}\,(t) = 2^{-j/2}\phi(2^{-j}t - k), \qquad j, k \in \mathbf{Z} \tag{1.18}$$

The basis wavelet functions are usually of the form

$$\psi_{j,k}\,(t) = 2^{-j/2}\psi(2^{-j}t - k), \qquad j, k \in \mathbf{Z} \tag{1.19}$$

It follows that there is a sequence $\{h_k\}$ (where h_k is a *scaling filter* associated with the wavelet) such that $\sum |h_k|^2 = 1$ and

$$f\,(t) = \sqrt{2} \sum_{k=0}^{\infty} h_k f\,(2t - k) \tag{1.20}$$

where ϕ is normalized so that $\int_{-\infty}^{\infty} \phi\,(t)\, dt = 1$.

When $\{h_k\}$ is finite, a compactly supported scaling function is the solution to the *dilation equation* above. The wavelet function is defined in terms of the scaling function as

$$\psi(t) = \sqrt{2} \sum_{k=0}^{\infty} g_k \phi(2t - k) \qquad (1.21)$$

where $\int_{-\infty}^{\infty} \psi(t)\, dt = 0$ and $g_k = (-1)^{k+1} h^{1-k}$ is a *wavelet filter*. Then $W_j = \text{span}\{\psi_{j,k}, k \in \mathbb{Z}\}$ is the orthogonal complement of V_j in V_{j+1}, $\forall j \in \mathbb{Z}$.

The DWT of the signal function comprises the *wavelet coefficients* $C(j, k)$, which are produced through the convulsion of a mother wavelet function $\psi_{j,k}(t)$ with the signal analyzed, $f(t)$:

$$C(j, k) = \int_{-\infty}^{+\infty} f(t)\psi_{j,k}(t)\, dt \qquad (1.22)$$

Thus, the discrete synthesis of the original signal is

$$f(t) = \sum_{j \in Z} \sum_{k \in Z} C(j, k)\psi_{j,k}(t) \qquad (1.23)$$

Usually, the low-frequency content is the most important part of a signal. It is what gives the signal its identity, while the high-frequency component usually contains a large part of the noise in the signal (Misiti et al., 2009). We term the high-scale, low-frequency components *approximations* and the low-scale, high-frequency components *details*. At each level j, we build the j-level approximation a_j, or approximation at level j, and a deviation signal called the j-level detail d_j, or detail at level j. The original signal is considered to be the approximation at level zero, denoted a_0. The words *approximation* and *detail* are justified by the fact that a_1 is an approximation of a_0 taking into account the low frequencies of a_0, whereas the detail d_1 corresponds to the high-frequency correction. For additional and detailed expositions on the mathematical aspects of wavelets, we refer, for example, to Daubechies (1992), Kaiser (1994), Kobayashi (1998), Mallat (1999), and Wojtaszczyk (1997).

Neural Networks

Next we present a brief introduction to neural networks. The advantages of neural networks are discussed as well as their applications in finance. In Chapter 2 the mathematical and theoretical aspects are presented analytically.

Artificial neural networks or, simply, neural networks draw their inspiration from biological systems. They are essentially devices of parallel and distributed processing for nonparametric statistical inference which simulate the basic operating principles of the biological brain. They consist of many interconnected neurones (also known as *hidden units*), whose associated weights determine the strength of the signal passed through them. No particular structure or parametric form is assumed a priori; rather, the strengths of the connections are computed in a way that captures the essential features in the data.

The iterative algorithms employed for this purpose are known as *learning algorithms* because of their gradual nature. Certain algorithms were firmly positioned in the statistical framework by White (1989) and later by Geman et al. (1992), who demonstrated that they could be formulated as a nonlinear regression problem.

Neural networks are analogous to nonparametric nonlinear regression models, which constitute a very powerful approach, especially for financial applications. More generally, they are essentially statistical tools for performing inductive inference, providing an elegant formalism for unifying different nonparametric paradigms, such as nearest neighbors, kernel smoothers, and projection pursuit. Neural networks have shown considerable successes in a variety of disciplines, ranging through engineering, control, and financial modeling (Zapranis and Refenes, 1999).

In the past 20 years, neural networks have enjoyed a rapid expansion and increasing popularity in both the academic and industrial research communities. Lately, an increased number of studies have been published in various research areas. The main power of neural networks accrues from their capability for universal function approximation. Many studies (e.g., Cardaliaguet and Euvrard, 1992; Cybenko, 1989; Funahashi, 1989; Hornik et al., 1989, 1990; Ito, 1993; White, 1990); have shown that one-hidden-layer neural networks can approximate arbitrarily well any continuous function, including the function's derivatives.

The novelty about neural networks lies in their ability to model nonlinear processes with few (if any) a priori assumptions about the nature of the generating process (Rumelhart et al., 1986). This is particularly useful in investment management, where much is assumed and little is known about the nature of the process determining asset prices.

Modern investment management models, such as the arbitrage pricing theory, rely on the assumption that asset returns can be explained in terms of a set of factors. The usual assumption is that the return of an asset is a linear combination of the asset's exposure to these factors. Such theories have been very useful in expanding our understanding of capital markets, but many financial anomalies have remained unexplainable. In Zapranis and Refenes (1999) the problem of investment management was divided into three *parts—factor analysis, estimating returns,* and *portfolio optimization*—and it was shown that neural learning can play a part in each.

Moreover, neural networks are less sensitive than classical approaches to assumptions regarding the error term, and hence they can be used in the presence of noise, chaotic sections, and probability distributions with fat tails. The usual assumptions among financial economists are that price fluctuations not due to external influences are dominated by noise and can be modeled by stochastic processes. As a result, analysts try to understand the nature of noise and develop tools for predicting its effects on asset prices. However, very often, these remaining price fluctuations are due to *nonlinear* dynamics of the generating process. Therefore, given appropriate tools, it is possible to represent (and possibly understand) more of the market's price structure on the basis of completely or partially deterministic but nonlinear dynamics (Zapranis and Refenes, 1999).

Because of their inductive nature, neural networks have the ability to infer complex nonlinear relationships between an asset price and its determinants. Although, potentially, this approach can lead to better nonparametric estimators, neural networks are not always easily accepted in the financial economics community, mainly because established procedures for testing the statistical significance of the various aspects of the estimated model do not exist. A coherent set of methodologies for developing and assessing neural models, with a strong emphasis on their practical use in the capital markets, was provided by Zapranis and Refenes (1999).

Recently, academic researchers and market participants, have shown an increasing interest in nonlinear modeling techniques, with neural networks assuming a prominent role. Neural networks are being applied to a number of "live" systems in financial engineering and have shown promising results. Various performance figures are being quoted to support these claims, but the absence of explicit models, due to the nonparametric nature of the approach, makes it difficult to assess the significance of the model estimated and the possibility that any short-term success is due to data mining (Zapranis and Refenes, 1999).

Due to their nonparametric nature, neural models are of particular interest in finance, since the deterministic part (if any) of asset price fluctuations is largely unknown and arguably nonlinear. Over the years they have been applied extensively to all stages of investment management. Comprehensive reviews have been provided by Trippi and Turban (1992), and by Vellido et al. (1999).

Among the numerous contributions, notable are applications in bonds by Moody and Utans (1994) and Dutta and Shekhar (1988); in stocks by White (1988), Refenes et al. (1994), and Schöneburg (1990); in foreign exchange by Weigend et al. (1991); in corporate and macroeconomics by Sen et al. (1992); and in credit risk by Atiya (2001).

From a statistical viewpoint, neural networks have a simple interpretation: Given a sample $D_n = \left\{ \left(\mathbf{x}_i; y_i \right) \right\}_{i=1}^{n}$ generated by an unknown function $\varphi(x)$ with the addition of a stochastic zero-mean component ε,

$$y_i = \varphi\left(\mathbf{x}_i\right) + \varepsilon_i \qquad (1.24)$$

the task of "neural learning" is to construct an estimator $g(\mathbf{x}; \mathbf{w}) \equiv \hat{\varphi}(\mathbf{x})$ of $\varphi(\mathbf{x})$, where $\mathbf{w}$ is a set of free parameters known as *connection weights*. Since no a priori assumptions are made regarding the functional form of $\varphi(\mathbf{x})$, the neural network $g(\mathbf{x}; \mathbf{w})$ is a *nonparametric* estimator of the conditional density $E[y|\mathbf{x}]$, as opposed to a *parametric* estimator, where a priori assumptions are made.

Nevertheless, knowledge of the relative theory of neural learning and the basic models of neural networks is essential for their effective implementation, especially in complex applications such as the ones encountered in finance. In Chapter 2 an in-depth analysis of neural networks is presented.

Wavelet Neural Networks

Wavelet networks are a new class of networks that combine classic sigmoid neural networks and wavelet analysis. Wavelet networks have been used with great success in a wide range of applications. Wavelet analysis has proved to be a valuable tool for analyzing a wide range of time series and has already been used with success in image processing, signal denoising, density estimation, signal and image compression, and time-scale decomposition. Wavelet analysis is often regarded as a "microscope" in mathematics (Cao et al., 1995), and it is a powerful tool for representing nonlinearities (Fang and Chow, 2006). The major drawback of wavelet analysis is that it is limited to applications of small input dimension. The reason for this is that the construction of a wavelet basis is computationally expensive when the dimensionality of the input vector is relatively high (Zhang, 1997).

On the other hand, neural networks have the ability to approximate any deterministic nonlinear process, with little knowledge and no assumptions regarding the nature of the process. However, classical sigmoid neural networks have a series of drawbacks. Typically, the initial values of the neural network's weights are chosen randomly. Random weight initialization is generally accompanied by extended training times. In addition, when the transfer function is sigmoidal, there is always a significant change: that the training algorithm will converge to local minima. Finally, there is no theoretical link between the specific parameterization of a sigmoidal activation function and the optimal network architecture, that is, model complexity (the opposite holds true for wavelet networks).

Pati and Krishnaprasad (1993) have demonstrated that it is possible to construct a theoretical formulation of a feedforward neural network in terms of wavelet decompositions. WNs were proposed by Zhang and Benveniste (1992) as an alternative to feedforward neural networks which would alleviate the aforementioned weaknesses associated with each method. The wavelet networks are a generalization of radial basis function networks. Wavelet networks are one hidden layer networks that use a wavelet as an activation function instead of the classic sigmoidal family. It is important here to mention that multidimensional wavelets preserve the "universal approximation" property that characterizes neural networks. The nodes (or wavelons) of wavelet networks are wavelet coefficients of the function expansion that have a significant value. Various reasons were presented by Bernard et al. (1998) explaining why wavelets should be used instead of other transfer functions. In particular, first, wavelets have strong compression abilities, and second, computing the value at a single point or updating the function estimate from a new local measure involves only a small subset of coefficients.

Wavelet networks have been used in a variety of applications so far: in short-term load forecasting (Bashir and El-Hawary, 2000; Benaouda et al., 2006; Gao and Tsoukalas, 2001; Ulugammai et al., 2007; Yao et al., 2000); in time-series prediction (Cao et al., 1995; Chen et al., 2006; Cristea et al., 2000); in signal classification and compression (Kadambe and Srinivasan, 2006; Pittner et al., 1998; Subasi et al., 2005); in signal denoising (Zhang, 2007); in statics, and dynamics (Allingham et al., 1998; Oussar and Dreyfus, 2000; Oussar et al., 1998; Pati and Krishnaprasad, 1993; Postalcioglu and Becerikli, 2007; Zhang and Benveniste, 1992); in nonlinear modeling (Billings and Wei, 2005); and in nonlinear static function approximation (Jiao et al., 2001; Szu et al., 1992; Wong and Leung, 1998). Khayamian et al. (2005) even proposed the use of wavelet networks as a multivariate calibration method for simultaneous determination of test samples of copper, iron, and aluminum. Finally,

Alexandridis and Zapranis (2013a,b) used wavelet networks in modeling and pricing financial weather derivatives.

In contrast to classical sigmoid neural networks, wavelet networks allow for constructive procedures that efficiently initialize the parameters of a network. Using wavelet decomposition a wavelet library can be constructed. In turn, each wavelon can be constructed using the best wavelet in the wavelet library. As a result, wavelet networks provide information for the relative participation of each wavelon to the function approximation and the estimated dynamics of the generating process. The main characteristics of these constructive procedures are (1) convergence to the global minimum of the cost function, and (2) the initial weight vector being in close proximity to the global minimum, resulting in drastically reduced training times (Zhang, 1997; Zhang and Benveniste, 1992). Finally, efficient initialization methods will approximate the same vector of weights, minimizing the loss function each time.

APPLICATIONS OF WAVELET NEURAL NETWORKS IN FINANCIAL ENGINEERING, CHAOS, AND CLASSIFICATION

Artificial intelligence and machine learning are used to improve the process of decision making in a wide range of financial applications. For example, they are used to evaluate credit risk, for risk assessment of mortgage loans, in project management and bidding strategies, in financial and economic forecasting, in risk assessment of investments in fixed-income products traded on an exchange, to identify regularities in price changes of securities, and to predict default risk. Other applications explored recently are portfolio selection and diversification, simulation of market behavior, and pattern identification in financial data.

During recent years, neural networks have been utilized extensively in various fields in finance. For example, Moody and Utans (1994) employed neural network in corporate bond rating prediction. Neural networks were used in securities trading by Trippi and Turban (1992) and Zapranis and Refenes (1999), and Chen et al. (2003) used them to forecast and trade the Taiwan stock index. Finally, neural networks were utilized for bankruptcy prediction by Tam and Kiang (1992) and Wilson and Sharda (1994).

On the other hand, wavelet networks have been introduced only recently in financial applications. As a result, the literature on the application of wavelet network in finance is limited. As an intermediate step, a very popular technique is to apply wavelet analysis to decompose a financial time series and then use the decomposed part or the wavelet coefficients as inputs to a neural network.

Becerra et al. (2005) used wavelet networks for the identification and prediction of corporate financial distress. More precisely, the authors used various financial ratios, such as working capital/total assets, accumulated retained profit/total assets, profit before interest, and tax/total assets. They collected data between 1997 and 2000 from 60 British firms and compared three models. The first one was a linear model; the other two were nonlinear nonparametric neural networks and wavelet networks. Their results indicate that the wavelet networks outperform the other two alternatives in classifying corporate financial distress correctly.

In a similar application wavelet networks were trained in predicting bankruptcy (Chauhan et al., 2009). Three data sets were used to test their model. The wavelet network was evaluated as to bankruptcy prediction for 129 U.S. banks, 66 Spanish banks, and 40 Turkish banks. Their results show high accuracy and conclude that wavelet networks are a very effective tool for classification problems.

Echauz and Vachtsevanos (1997) used wavelet networks as trading advisors. Based on past data, the investor had to decide whether to switch in or out of an S&P index fund. The decisions suggested by the wavelet networks were based on a 13-week holding period yield. Their results indicated that wavelet networks exceeded the performance of a simple buy-and-hold strategy.

Wavelet analysis was used by Aussem et al. (1998) to decompose the S&P index. The daily price index was decomposed on various scales, and then the coefficients of the wavelet decompositions were used as the input to a network to forecast the future price of the index up to five days ahead.

Similarly, wavelet analysis and radial basis function neural networks (a subclass of wavelet networks) were combined by Liu et al. (2004). The two methods were combined to identify and analyze possible chart patterns, and a method of financial forecasting using technical analysis was presented. Their results indicate that the matching method proposed can be used to automate the chart pattern-matching process.

Bashir and El-Hawary (2000) applied a wavelet network to short-term load forecasting. Similarly, Benaouda et al. (2006), Gao and Tsoukalas (2001), and Ulugammai et al. (2007) utilized different versions of wavelet networks and wavelet analysis for accurate forecasting of electricity loads, a necessity in the management of energy companies. Pindoriya et al. (2008) utilized wavelet networks in energy price forecasting in electricity markets.

Wavelet networks have also been used to analyze financial time series (Alexandridis and Zapranis, 2013a,b; Zapranis and Alexandridis, 2011). More precisely, wavelet networks were employed in the context of weather derivative pricing and modeling. Moreover, trained wavelet networks were used to predict the future prices of financial weather derivatives. Their findings suggest that wavelet networks can model the underlying processes very well and thus constitute a very accurate and efficient tool for weather derivative pricing.

Wavelet networks were applied to the oil market by Alexandridis et al. (2008). The objective was to investigate the factors that affect crude oil price for the period 1988 to 2008 through comparison of a linear and a nonlinear approach. The dynamics of the West Texas Intermediate crude oil price, time series as well as the relation of crude oil price, and returns to various explanatory variables were studied. First, wavelet analysis was used to extract the driving forces and dynamics of crude oil price and returns processes. Also examined was whether a wavelet neural network estimator could provide some incremental value toward understanding the crude oil price process. In contrast to the linear model, it was determined that wavelet network findings are in line with those of economic analysts. Wavelet networks not only have better predictive power in forecasting crude oil returns but can also model the dynamics of the returns correctly.

Alexandridis and Hasan (2013) investigated the impact of the global financial crisis on the multihorizon nature of systematic risk and market risk using daily data from eight major European equity markets over the period 2005 to 2012. The method was based on a wavelet multiscale approach within the framework of a capital asset pricing model. The sample covers pre-crisis, crisis, and post-crisis periods with varying experiences and regimes. The results indicate that beta coefficients have a mutliscale tendency in sample countries. Moreover, wavelet networks were trained using past data to forecast the post-crisis multiscale betas and the value at risk in each market.

In addition, wavelet networks have been employed extensively in classification problems in various disciplines other than finance: for example, in medicine in recognition of cardiac arythmias (Lin et al., 2008) and breast cancer (Senapati et al., 2011) and for EEG signal classification (Subasi et al., 2005); by Szu et al. (1992) for signal classification and by Shi et al. (2006) for speech signal processing. Finally, wavelet networks were utilized successfully for the prediction of chaotic time series (Cao et al., 1995; Chen et al., 2006; Cristea et al., 2000).

BUILDING WAVELET NETWORKS

In this book we focus on the construction of optimal wavelet networks and their successful application in a wide range of applications. A generally accepted framework for applying wavelet networks is missing from the literature. In this book we present a complete statistical model identification framework in order to employ wavelet networks in various applications. To our knowledge, we are the first to do so. Although a vast literature on wavelet networks exists, to our knowledge this is the first study that presents a step-by-step guide for model identification for wavelet networks. Model identification can be classified in two parts: model selection and variable significance testing. In this study, a framework similar to the one proposed by Zapranis and Refenes (1999) for classical sigmoid neural networks is adapted. More precisely, the following subjects were examined thoroughly: the structure of a wavelet network, training methods, initialization algorithms, variable significance and variable selection algorithms, model selection methods, and methods to construct confidence and prediction intervals. Some of these issues were studied to some extent by Iyengar et al. (2002).

Moreover, to succeed in the use of wavelet networks, a model adequacy framework is needed. The evaluation of the model is usually performed in two stages. In the first the accuracy of the predictions is assessed and in the second the performance and behavior of the wavelet network are evaluated under conditions as close as possible to the actual operating conditions of the application.

Variable Selection

In the following chapters, the first part of model identification, variable selection, is discussed analytically. The objective is to develop an algorithm to use to select the statistically significant variables from a group of candidates in a problem where little theory is available. Our aim is to present a statistical procedure that produces robust and stable results when it is applied in the wavelet network framework.

In real problems it is important to determine the independent variables correctly, for various reasons. In most financial problems there is little theoretical guidance and information about the relationship of any explanatory variable with the dependent variable. Irrelevant variables are among the most common sources of specification error. An unnecessary independent variable included in a model (e.g., in the training of a wavelet network) creates a series of problems. First, the architecture of the wavelet network increases. As a result, the training time of the wavelet network and the computational burden can increase significantly. Moreover, there is a possibility that the inclusion of irrelevant variables will affect the training of the wavelet network. As presented for complex problems in subsequent chapters, when irrelevant variables are included there is a possibility that the training algorithm will be trapped in a local minimum during minimization of the loss function. Finally, when irrelevant variables are included in the model, its predictive power and generalization ability are reduced.

On the other hand, correctly specified models are easier to understand and interpret. The underlying relationship is explained by only a few key variables, while all minor and random influences are attributed to the error term. Finally, including a large number of independent variables relative to sample size runs the risk of a spurious fit.

To select the statistically significant and relevant variable, from a group of possible explanatory variables, hypotheses tests of statistical significance are used. To do so, the relative contributions of the explanatory variables in explaining the dependent variable in the context of a specific model are estimated. Then the significance of a variable is assessed. This can be done by testing the null hypothesis that it is irrelevant, either by comparing a test statistic that has a known theoretical distribution with its critical value, or by constructing confidence intervals for the relevance criterion. Our decision on rejecting or not rejecting the null hypothesis is based on a given significance level. The p-value, the smallest significance level for which the null hypothesis will not be refuted, imposes a ranking on the variables according to their relevance to the particular model (Zapranis and Refenes, 1999).

Before proceeding to variable selection, a measure of relevance must be defined. Various measures are tested in later chapters, and an algorithm that produces robust and stable results is presented.

Model Selection

Next, the second part of model identification, variable selection, will be discussed analytically. The objective is to find a statistical procedure that identifies correctly the optimal number of wavelons (hidden units) that are needed to model a specific problem. One of the most crucial steps is to identify the correct topology of the network. A desired wavelet network architecture should contain as few hidden units as necessary while explaining as much variability of the training data as possible. A network with fewer hidden units than needed would not be able to learn the underlying function, while selecting more hidden units than needed will result in an overfitted model.

In both cases, the results obtained from the wavelet network cannot be interpreted. Moreover, an underfitted or overfitted network cannot be used to forecast the evolution of the underlying process. In the first case the model has not learned the dynamics of the underlying process, hence cannot be used for forecasting. On the other hand, in an overfitted network, the model has learned the noise part; hence, any forecasts will be affected by the noise and will differ significantly from the real target values. Therefore, an algorithm to select the appropriate wavelet network model for a given problem must be derived.

Model Adequacy Testing

After the construction of a wavelet network, we are interested in measuring its predictive ability in the context of a particular application. The evaluation of the model usually includes two clearly distinct, though related stages. In the first stage, various metrics that quantify the accuracy of the predictions or the classifications made by the model are used and the model is evaluated based on these metrics. The term *accuracy* is a quantification of the "proximity" between the outputs of the wavelet network and the target values desired. Measurements of the precision are related to the error function that is minimized (or in some cases, the profit function that is maximized) during the model specification of the wavelet network model.

The second step is to assess the behavior and performance of the wavelet network model under conditions as close as possible to the actual operating conditions of the application. The greater accuracy of the network model does not necessarily mean that it will be applied successfully. It is important, therefore, that the performance of the model be evaluated in the context of the decision-making system that it supports. Especially for use in time-series forecasting application, the performance and evaluation of the model should be based on benchmark trading strategies. It is also important to evaluate the behavior of the model throughout the range of the actual scenarios possible. For example, if the application concerns the prediction of the performance of a stock index, it would not be correct if the validation sample corresponds solely to a period with a strong upward trend, since the evaluation will be restricted to the specific circumstances.

A full understanding of the behavior of the model under a variety of conditions is a prerequisite for the creation of a successful decision support system or simply a decision-making system (e.g., automated trading systems). In upcoming chapters, an analytical model adequacy framework is presented.

BOOK OUTLINE

The purpose of the book is twofold: first, to expand the framework that was developed by Alexandridis (2010) for model selection and variable selection in the framework of wavelet networks; and second, to provide a textbook that presents a step-by-step guide to employing wavelet networks in various applications (finance, classification, chaos, etc.).

Chapter 2: Neural Networks

In this chapter the basic aspects of the neural networks are presented: more precisely, the delta rule, the backpropagation algorithm, and the concept of training a network. Our purpose is to make the reader familiar with the basic concepts of the neural network framework. These concepts are later modified to construct a new class of neural networks called wavelet networks.

Chapter 3: Wavelet Neural Networks

The basic aspects of the wavelet networks are presented in this chapter: more precisely, the structure of a wavelet network, various initialization methods, a training method for the wavelet networks, and stopping conditions of the training. In addition, online training is discussed and various initialization methods are compared and evaluated in two case studies.

Chapter 4: Model Selection: Selecting the Architecture of the Network

Chapter 4 we describe the model selection procedure. One of the most crucial steps is to identify the correct topology of the network. Various algorithms and criteria are presented as well as a complete statistical framework. Finally, the various methods are compared and evaluated in two case studies.

Chapter 5: Variable Selection: Determining the Explanatory Variables

In this chapter various methods of testing the significance of each explanatory variable are presented and tested. The purpose of this section is to find an algorithm that gives consistently stable and correct results when it is used with wavelet networks. In real problems it is important to determine correctly the independent variables. In most problems there is only limited information about the relationship of any explanatory variable with the dependent variable. As a result, unnecessary independent variables are included in the model, reducing its predictive power. Finally, the various methods are compared and evaluated in two case studies.

Chapter 6: Model Adequacy Testing: Determining a Network's Future Performance

In this chapter we present various methods used to test the adequacy of the wavelet network constructed. Various criteria are presented that examine the residuals of the wavelet network. Moreover, depending on the application (classification or prediction), additional evaluation criteria are presented and discussed.

Chapter 7: Modeling Uncertainty: From Point Estimates to Prediction Intervals

The framework proposed is expanded by presenting two methods of estimating confidence and prediction intervals. The output of the wavelet networks is the approximation of the underlying function obtained from noisy data. In many applications, especially in finance, risk managers may be more interested in predicting intervals for future movements of the underlying function than in simply point estimates. In addition, the methods proposed are compared and evaluated in two case studies.

Chapter 8: Modeling Financial Temperature Derivatives

In this chapter a real data set is used to demonstrate the application of our proposed framework to a financial problem. More precisely, using data from detrended and deseasonalized daily average temperatures, a wavelet network is constructed, initialized, and trained, in the context of modeling and pricing financial weather derivatives. At the same time the significant variables are selected: in this case, the correct number of lags. Finally, the wavelet network developed will be used to construct confidence and prediction intervals.

Chapter 9: Modeling Financial Wind Derivatives

In this chapter, the framework proposed is applied to a second financial data set. More precisely, daily average wind speeds are modeled and forecast in the context of modeling and pricing financial weather derivatives. At the same time, the significant variables are selected: in this case, the correct number of lags. Finally, the wavelet network developed will be used to forecast the future index of the wind derivatives.

Chapter 10: Predicting Chaotic Time Series

In this chapter the framework proposed is evaluated for modeling and predicting chaotic time series. More precisely, the chaotic system is described by the Mackey–Glass equation. A wavelet network is constructed and then used to predict the evolution of the chaotic system and construct confidence and prediction intervals.

Chapter 11: Classification of Breast Cancer Cases

In this chapter the framework proposed in the earlier chapters is applied in a real-life application. In this case study a wavelet network is constructed to classify breast cancer based on various attributes. Each instance has one of two possible classes: benign or malignant. The aim is to construct a wavelet network that accurately classifies each clinical case. The classification is based on nine attributes: clump thickness, uniformity of cell size, uniformity of cell shape, marginal adhesion, single epithelial cell size, bare nuclei, bland chromatin, normal nucleoli, and mitoses.

REFERENCES

Alexandridis, A. (2010). "Modelling and pricing temperature derivatives using wavelet networks and wavelet analysis," University of Macedonia, Thessaloniki, Greece.

Alexandridis, A., and Hasan, M. (2013). "Global financial crisis and multiscale systematic risk: evidence from selected european markets." In *The Impact of the Global Financial Crisis: on Banks, Financial Markets and Institutions in Europe*, University of Southampton, UK.

Alexandridis, A., and Livanis, S. (2008). "Forecasting crude oil prices using wavelet neural networks." *5th ΦΣΔΕΤ* Athens, Greece.

Alexandridis, A., and Zapranis, A. (2013a). "Wind derivatives: modeling and pricing." *Computational Economics*, 41(3), 299–326.

Alexandridis, A. K., and Zapranis, A. D. (2013b). *Weather Derivatives: Modeling and Pricing Weather-Related Risk*. Springer-Verlag, New York.

Alexandridis, A., Zapranis, A., and Livanis, S. (2008). "Analyzing crude oil prices and returns using wavelet analysis and wavelet networks." *7th HFAA*, Chania, Greece.

Allen, J. B. (1982). "Application of the short-time Fourier transform to speech processing and spectral analysis." *IEEE ICASSP*, 1012–1015.

Allingham, D., West, M., and Mees, A. I. (1998). "Wavelet reconstruction of nonlinear dynamics." *International Journal of Bifurcation and Chaos*, 8(11), 2191–2201.

Atiya, A. F. (2001). "Bankruptcy prediction for credit risk using neural networks: A survey and new results." *IEEE Transactions on Neural Networks*, 12(4), 929–935.

Aussem, A., Campbell, J., and Murtagh, F. (1998). "Wavelet-based feature extraction and decomposition strategies for financial forecasting." *Journal of Computational Intelligence in Finance*, 6, 5–12.

Bashir, Z., and El-Hawary, M. E. (2000). "Short term load forecasting by using wavelet neural networks." *Proceedings of Canadian Conference on Electrical and Computer Engineering*, 163–166.

Becerra, V. M., Galvão, R. K., and Abou-Seada, M. (2005). "Neural and wavelet network models for financial distress classification." *Data Mining and Knowledge Discovery*, 11(1), 35–55.

Benaouda, D., Murtagh, G., Starck, J.-L., and Renaud, O. (2006). "Wavelet-based nonlinear multiscale decomposition model for electricity load forecasting." *Neurocomputing*, 70, 139–154.

Bernard, C., Mallat, S., and Slotine, J.-J. (1998). "Wavelet interpolation networks." *Proceedings of ESANN '98*, Bruges, Belgium, 47–52.

Billings, S. A., and Wei, H.-L. (2005). "A new class of wavelet networks for nonlinear system identification." *IEEE Transactions on Neural Networks*, 16(4), 862–874.

Black, F., and Scholes, M. (1973). "The pricing of options and corporate liabilities." *Journal of Political Economy*, 81(3), 637–654.

Bracewell, R. N. (2000). *The Fourier Transform and Its Applications*. McGraw-Hill, New York.

Cao, L., Hong, Y., Fang, H., and He, G. (1995). "Predicting chaotic time series with wavelet networks." *Physica*, D85, 225–238.

Cardaliaguet, P., and Euvrard, G. (1992). "Approximation of a function and its derivative with a neural network." *Neural Networks*, 5(2), 207–220.

Chauhan, N., Ravi, V., and Karthik Chandra, D. (2009). "Differential evolution trained wavelet neural networks: application to bankruptcy prediction in banks." *Expert Systems with Applications*, 36(4), 7659–7665.

Chen, A.-S., Leung, M. T., and Daouk, H. (2003). "Application of neural networks to an emerging financial market: forecasting and trading the Taiwan stock index." *Computers and Operations Research*, 30(6), 901–923.

Chen, Y., Yang, B., and Dong, J. (2006). "Time-series prediction using a local linear wavelet neural wavelet." *Neurocomputing*, 69, 449–465.

Cristea, P., Tuduce, R., and Cristea, A. (2000). "Time series prediction with wavelet neural networks." *Proceedings of 5th Seminar on Neural Network Applications in Electrical Engineering*, Belgrade, Yugoslavia, 5–10.

Cybenko, G. (1989). "Approximation by superpositions of a sigmoidal function." *Mathematics of Control, Signals and Systems*, 2(4), 303–314.

Daubechies, I. (1992). *Ten Lectures on Wavelets*. SIAM, Philadelphia.

Donoho, D. L., and Johnstone, I. M. (1994). "Ideal spatial adaption by wavelet shrinkage." *Biometrika*, 81, 425–455.

Dutta, S., and Shekhar, S. (1988). "Bond rating: a nonconservative application of neural networks." *IEEE International Conference on Neural Networks*, 443–450.

Echauz, J., and Vachtsevanos, G. (1997). "Separating order from disorder in a stock index using wavelet neural networks." *EUFIT*, 97, 8–11.

Fang, Y., and Chow, T. W. S. (2006). "Wavelets based neural network for function approxima-tion." *Lecture Notes in Computer Science*, 3971, 80–85.

Fernandez, V. P. (2005). "The international CAPM and a wavelet-based decomposition of value at risk." *Studies in Nonlinear Dynamics and Econometrics*, 9(4).

Fernandez, V. (2006). "The CAPM and value at risk at different time-scales." *International Review of Financial Analysis*, 15(3), 203–219.

Finnerty, J. D. (1988). "Financial engineering in corporate finance: an overview." *Financial Management*, 14–33.

Funahashi, K.-I. (1989). "On the approximate realization of continuous mappings by neural Networks." *Neural networks*, 2(3), 183–192.

Gao, R., and Tsoukalas, H. I. (2001). "Neural-wavelet methodology for load forecasting." *Journal of Intelligent and Robotic Systems*, 31, 149–157.

Geman, S., Bienenstock, E., and Doursat, R. (1992). "Neural networks and the bias/variance dilemma." *Neural Computation*, 4(1), 1–58.

Gençay, R., Selçuk, F., and Whitcher, B. (2003). "Systematic risk and timescales." *Quantitative Finance*, 3, 108–116.

Gençay, R., Selçuk, F., and Whitcher, B. (2005). "Multiscale systematic risk." *Journal of International Money and Finance*, 24(1), 55–70.

Grossmann, A., and Morlet, J. (1984). "Decomposition of Hardy functions intro square-integrable wavelets of constant shape." *SIAM Journal of Mathematical Analysis*, 15(4), 723–736.

Hawley, D. D., Johnson, J. D., and Raina, D. (1990). "Artificial neural systems: a new tool for financial decision-making." *Financial Analysts Journal*, 63–72.

He, K., Lai, K. K., and Yen, J. (2012). "Ensemble forecasting of value at risk via multi resolution analysis based methodology in metals markets." *Expert Systems with Applications*, 39(4), 4258–4267.

Hornik, K., Stinchcombe, M., and White, H. (1989). "Multilayer feedforward networks are universal approximators." *Neural Networks*, 2(5), 359–366.

Hornik, K., Stinchcombe, M., and White, H. (1990). "Universal approximation of an unknown mapping and its derivatives using multilayer feedforward networks." *Neural Networks*, 3(5), 551–560.

Huang, C.-f., and Litzenberger, R. H. (1988). *Foundations for Financial Economics*. North-Holland, New York.

In, F., and Kim, S. (2006a). "The hedge ratio and the empirical relationship between the stock and futures markets: a new approach using wavelet analysis." *Journal of Business*, 79(2), 799–820.

In, F., and Kim, S. (2006b). "Multiscale hedge ratio between the Australian stock and futures markets: Evidence from wavelet analysis." *Journal of Multinational Financial Management*, 16(4), 411–423.

In, F., and Kim, S. (2007). "A note on the relationship between Fama–French risk factors and innovations of ICAPM state variables." *Finance Research Letters*, 4(3), 165–171.

Ito, Y. (1993). "Extension of approximation capability of three layered neural networks to derivatives." *IEEE International Conference on Neural Networks*, pp. 377–381.

Iyengar, S. S., Cho, E. C., and Phoha, V. V. (2002). *Foundations of Wavelet Networks and Applications*. CRC Press, Grand Rapids, MI.

Jiao, L., Pan, J., and Fang, Y. (2001). "Multiwavelet neural network and its approximation properties." *IEEE Transactions on Neural Networks*, 12(5), 1060–1066.

Kadambe, S., and Srinivasan, P. (2006). "Adaptive wavelets for signal classification and compression." *International Journal of Electronics and Communications*, 60, 45–55.

Kaiser, G. (1994). *A Friendly Guide To Wavelets*. Birkhauser, Cambridge, MA.

Khayamian, T., Ensafi, A. A., Tabaraki, R., and Esteki, M. (2005). "Principal component-wavelet networks as a new multivariate calibration model." *Analytical Letters*, 38(9), 1447–1489.

Kim, S., and In, F. (2005). "The relationship between stock returns and inflation: new evidence from wavelet analysis." *Journal of Empirical Finance*, 12, 435–444.

Kim, S., and In, F. (2007). "On the relationship between changes in stock prices and bond yields in the G7 countries: wavelet analysis." *Journal of International Financial Markets, Money and Finance*, 17, 167–179.

Kobayashi, M. (1998). *Wavelets and Their Applications*. SIAM, Philadelphia.

Lim, J. S. (1990). *Two-Dimensional Signal And Image Processing*. Prentice Hall, Englewood Cliffs, NJ.

Lin, C.-H., Du, Y.-C., and Chen, T. (2008). "Adaptive wavelet network for multiple cardiac arrhythmias recognition." *Expert Systems with Applications*, 34(4), 2601–2611.

Liu, J. N. K., Kwong, R. W. M., and Bo, F. (2004). "Chart patterns recognition and forecast using wavelet and radial basis function networks." *Lecture Notes in Computer Science*, 3214, 564–571.

Maharaj, E. A., Galagedera, D. U. A., and Dark, J. (2011). "A comparison of developed and emerging equity market return volatility at different time scales." *Managerial Finance*, 37(10), 940–952.

Mallat, S. G. (1999). *A Wavelet Tour of Signal Processing*. Academic Press, San Diego, CA.

Marshall, and Bansal, V. K. (1992). *Financial Engineering: A Complete Guide to Financial Innovation*. New York Institute of Finance, New York.

Masih, M., Alzahrani, M., and Al-Titi, O. (2010). "Systematic risk and time scales: new evidence from an application of wavelet approach to the emerging Gulf stock markets." *International Review of Financial Analysis*, 19(1), 10–18.

Mason, S. P., Merton, R. C., Perold, A. F., and Tufano, P. (1995). *Cases in Financial Engineering: Applied Studies of Financial Innovation*, Prentice Hall, Englewood Cliffs, NJ.

Misiti, M., Misiti, Y., Oppenheim, G., and Poggi, J.-M. (2009). *Wavelet Toolbox 4: User's Guide*. MathWorks.

Moody, J. E., and Utans, J. (1994). "Architecture selection strategies for neural networks: Application to corporate bond rating prediction." In *Neural Networks in the Capital Markets*, A. P. Refenes, ed. Wiley, New York.

Mulvey, J. M., Rosenbaum, D. P., and Shetty, B. (1997). "Strategic financial risk management and operations research." *European Journal of Operational Research*, 97(1), 1–16.

Nawab, S., Quatieri, T., and Lim, J. (1983). "Signal reconstruction from short-time Fourier transform magnitude." *IEEE Transactions on Acoustics, Speech, Signal Processing*, 31, 986–998.

Norsworthy, J. R., Li, D., and Gorener, R. (2000). "Wavelet-based analysis of time series: an export from engineering to finance." *Proceedings of the 2000 IEEE*, 126–132.

Oppenheim, A., Schafer, R., and Buck, J. (1999). *Discrete Time Signal Processing.* Prentice Hall, Englewood Cliffs, NJ.

Oussar, Y., and Dreyfus, G. (2000). "Initialization by selection for wavelet network training." *Neurocomputing*, 34, 131–143.

Oussar, Y., Rivals, I., Presonnaz, L., and Dreyfus, G. (1998). "Trainning wavelet networks for nonlinear dynamic input output modelling." *Neurocomputing*, 20, 173–188.

Pati, Y. C., and Krishnaprasad, P. S. (1993). "Analysis and synthesis of feedforward neural networks using discrete affine wavelet transforms." *IEEE Transactions on Neural Networks*, 4(1), 73–85.

Pindoriya, N., Singh, S., and Singh, S. (2008). "An adaptive wavelet neural network-based energy price forecasting in electricity markets." *IEEE Transactions on Power Systems*, 23(3), 1423–1432.

Pittner, S., Kamarthi, S. V., and Gao, Q. (1998). "Wavelet networks for sensor signal classification in flank wear assessment." *Journal of Intelligent Manufacturing*, 9, 315–322.

Postalcioglu, S., and Becerikli, Y. (2007). "Wavelet networks for nonlinear system modelling." *Neural Computing and Applications*, 16, 434–441.

Rabeh, K. K., and Mohamed, B. B. (2011). "A time-scale analysis of systematic risk: wavelet-based approach." Munich Personal RePEc Archive, Paper No. 31938, http://mpra.ub.uni-muenchen.de/31938/

Ramsey, J. B. (1999). "The contribution of wavelets to the analysis of economic and financial data." *Philosophical Transactions: Mathematical, Physical and Engineering Sciences*, 357(1760), 2593–2606.

Refenes, A., Zapranis, A., and Francis, G. (1994). "Stock performance modeling using neural networks: a comparative study with regression models." *Neural Networks*, 7(2), 375–388.

Rua, A., and Nunes, L. C. (2012). "A wavelet-based assessment of market risk: the emerging markets case." *Quarterly Review of Economics and Finance*, 52(1), 84–92.

Rumelhart, D. E., McClelland, J. L., and University of California–San Diego PDP Research Group. (1986). *Parallel Distributed Processing: Explorations in the Microstructure of Cognition.* MIT Press, Cambridge, MA.

Schöneburg, E. (1990). "Stock price prediction using neural networks: a project report." *Neurocomputing*, 2(1), 17–27.

Sen, T., Oliver, R., and Sen, N. (1992). "Predicting Corporate Mergers Using Backpropagation Neural Networks: A Comparative Study with Logistic Models." Department of Accounting, RB Pamblin College of Business, Virginia Tech, Blacksburg, VA.

Senapati, M., Mohanty, A., Dash, S., and Dash, P. (2011). "Local linear wavelet neural network for breast cancer recognition." *Neural Computing and Applications*, 1–7.

Shi, D., Chen, F., Ng, G. S., and Gao, J. (2006). "The construction of wavelet network for speech signal processing." *Neural Computing and Applications*, 15, 217–222.

Subasi, A., Alkan, A., Koklukaya, E., and Kiymik, M. K. (2005). "Wavelet neural network classification of EEG signals by using AR model with MLE pre-processing." *Neural Networks*, 18, 985–997.

Szu, H., Telfer, B., and Kadambe, S. (1992). "Neural network adaptive wavelets for signal representation and classification." *Optical Engineering*, 31, 1907–1916.

Tam, K. Y., and Kiang, M. Y. (1992). "Managerial applications of neural networks: the case of bank failure predictions." *Management Science*, 38(7), 926–947.

Trippi, R. R., and Turban, E. (1992). *Neural Networks in Finance and Investing: Using Artificial Intelligence to Improve Real World Performance*. McGraw-Hill, New York.

Ulugammai, M., Venkatesh, P., Kannan, P. S., and Padhy, N. P. (2007). "Application of bacterial foraging technique trained artificial and wavelet neural networks in load forecasting." *Neurocomputing*, 70, 2659–2667.

Vellido, A., Lisboa, P. J., and Vaughan, J. (1999). "Neural networks in business: a survey of applications (1992–1998)." *Expert Systems with Applications*, 17(1), 51–70.

Walker, J. S. (2008). *A Primer on Wavelets and Their Scientific Applications*, Chapman and Hall, New York.

Weigend, A. S., Rumelhart, D. E., and Huberman, B. A. (1991). "Generalization by weight-elimination applied to currency exchange rate prediction." *IJCNN-91-Seattle International Joint Conference on Neural Networks*, 837–841.

White, H. (1988). "Economic prediction using neural networks: the case of IBM daily stock returns." *IEEE International Conference on Neural Networks*, 451–458.

White, H. (1989). "Learning in artificial neural networks: a statistical perspective." *Neural Computation*, 1, 425–464.

White, H. (1990). "Connectionist nonparametric regression: multilayer feedforward networks can learn arbitrary mappings." *Neural Networks*, 3(5), 535–549.

Wilson, R. G. (1995). *Fourier Series and Optical Transform Techniques in Contemporary Optics*. Wiley, New York.

Wilson, R. L., and Sharda, R. (1994). "Bankruptcy prediction using neural networks." *Decision Support Systems*, 11(5), 545–557.

Wojtaszczyk, P. (1997). *A Mathematical Introduction to Wavelets*. Cambridge University Press, Cambridge, UK.

Wong, K.-W., and Leung, A. C.-S. (1998). "On-line successive synthesis of wavelet networks." *Neural Processing Letters*, 7, 91–100.

Yao, S. J., Song, Y. H., Zhang, L. Z., and Cheng, X. Y. (2000). "Wavelet transform and neural networks for short-term electrical load forecasting." *Energy Conversion and Management*, 41, 1975–1988.

Yousefi, S., Weinreich, I., and Reinarz, D. (2005). "Wavelet-based prediction of oil prices." *Chaos, Solitons and Fractals*, 25(2), 265–275.

Zapranis, A., and Alexandridis, A. (2006). "Wavelet Analysis and Weather Derivatives Pricing." 5th Hellenic Finance and Accounting Association, Thessaloniki, Greece.

Zapranis, A., and Alexandridis, A. (2008). "Modelling temperature time dependent speed of mean reversion in the context of weather derivetive pricing." *Applied Mathematical Finance*, 15(4), 355–386.

Zapranis, A., and Alexandridis, A. (2009). "Weather derivatives pricing: modelling the seasonal residuals variance of an ornstein–uhlenbeck temperature process with neural networks." *Neurocomputing*, 73, 37–48.

Zapranis, A., and Alexandridis, A. (2011). "Modeling and forecasting cumulative average temperature and heating degree day indices for weather derivative pricing." *Neural Computing and Applications*, 20(6), 787–801.

Zapranis, A., and Refenes, A. P. (1999). Principles of Neural Model Indentification, Selection and Adequacy: With Applications to Financial Econometrics. Springer-Verlag, New York.

Zapranis, A., and Sivridis, S. (2003). "Extending Vasicek with neural regression." *Neural Network World*, 13(2), 187–210.

Zhang, Q. (1997). "Using wavelet network in nonparametric estimation." *IEEE Transactions on Neural Networks*, 8(2), 227–236.

Zhang, Q., and Benveniste, A. (1992). "Wavelet networks." *IEEE Transactions on Neural Networks*, 3(6), 889–898.

Zhang, Z. (2007). "Learning algorithm of wavelet network based on sampling theory." *Neurocomputing*, 71(1), 224–269.

Ζαπράνης, Α. (2005). *Χρηματοοικονομική και Νευρωνικά Συστήματα*, Κλειδάριθμος, Αθήνα.

2

Neural Networks

Artificial neural networks derive their origins from biological neural networks. They are systems of parallel and distributed processing that simulate the basic operating principles of the biological brain. Sometimes they are referred to in the literature as *machine learning algorithms*. The biological brain in its basic structure is a network of neural cells (neurons) attached through connections that have the ability to adjust the power of the electrical pulse that runs through them (synapses). The external stimulus in the form of an electrical pulse is transmitted as information through synapses to the neurons, where it is processed, and eventually, an output response of the network is produced. The information is encoded as "knowledge" through continuous updating of the existing synapses between neurons.

In this book we treat neural networks as the eminent expression of nonparametric regression. Nonparametric regression is a very powerful approach, especially for financial applications. Neural networks can approximate any unknown nonlinear function and are generally less sensitive than classical approaches to assumptions about the error term; hence, they can perform well in the presence of noise, chaotic sections, and fat tails of probability distributions.

The basic aspects of neural networks are presented below. More precisely, the usual training algorithm and network structures are presented. In addition, a geometric explanation of the backpropagation learning rule is given.

Wavelet Neural Networks: With Applications in Financial Engineering, Chaos, and Classification,
First Edition. Antonios K. Alexandridis and Achilleas D. Zapranis.
© 2014 John Wiley & Sons, Inc. Published 2014 by John Wiley & Sons, Inc.

PARALLEL PROCESSING

Although there are now many types of artificial neural networks, they all have one common characteristic: They are systems of parallel distributed processing (PDP). The processing of information is distributed over several computing units, while its encryption is accomplished by the interactions of all these units. Each PDP system consists of the following components (Rumelhart et al., 1986b):

- A set of processing units.
- A state of activation.
- An output function for each unit.
- An activation rule that combines all the inputs of a unit with the current activation state in order to compute the new activation state.
- A connectivity model between units.
- A signal propagation rule through the connections between units.
- A learning rule according to which the connectivity model alters through training.
- An environment in which the system must operate.

Usually, the output function and the activation rule are the same. A schematic representation of a PDP system is shown in Figure 2.1, and Figure 2.2 illustrates the general features of a processing unit. Below we examine the most important of these features and the assumptions made for them, and the interdependence of the components within a PDP system.

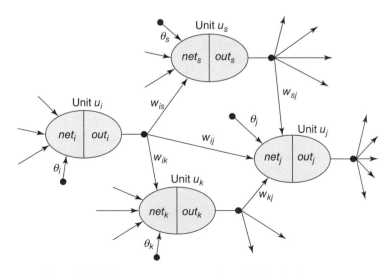

Figure 2.1 *Schematic representation of parallel distributed processing.*

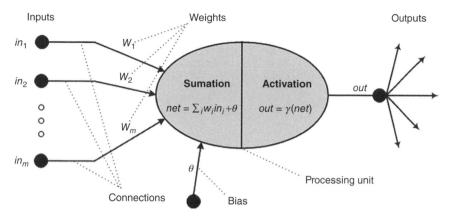

Figure 2.2 *Schematic representation of the general characteristics of a processing unit (an artificial neuron).*

Processing Units

Each processing unit receives inputs from the neighboring units or external sources and uses them to calculate an output signal which is distributed to other units. As illustrated in Figure 2.2, a unit receives some input signals, in_1, in_2, ..., in_m, which, unlike electrical pulses of the biological neuron, correspond to continuous variables. A weight value w_i changes any such input signal. The role of the weights is equivalent to the biological neuron's synapses. The value of the weight can be positive or negative. The input signal that is transferred via the bias connection is constant and has a value of 1.

The body of the artificial neuron is divided into two sections. In the first, the net input is estimated by the weighted summation of the inputs, *net*; in the second, the output value is estimated by the activation function γ. We can distinguish among three types of units: input units, which receive the data; output units, which are sending data out of the network; and hidden units, whose inputs and outputs signals are staying inside the network. During operation of the neural network, the units can be adjusted synchronously or asynchronously. In synchronous updating, all units update their activation level simultaneously; in asynchronous updating each unit has some (usually, fixed) probability to adjust its activation level at time t, and usually only one unit will be able to do so at this particular time.

Activation Status and Activation Rules

A rule that defines the effect of the net input to the activation state of a unit is needed. This rule is given by the function γ, which is also called an *activation function* or *transfer function*, and takes as inputs the net input net (t) and the current activation status out (t) at time t and returns as output the new value of the activation state out $(t + 1)$ at time $t + 1$:

$$\text{out}\,(t+1) = \gamma\,(\text{net}\,(t)\,,\text{out}\,(t)) \qquad (2.1)$$

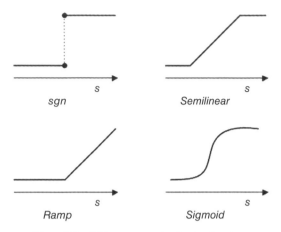

Figure 2.3   *Different types of activation functions.*

Usually, the activation function is a nondecreasing function of the net input only. Also, the activation functions are not strictly limited to nondecreasing forms. That is,

$$\text{out}(t+1) = \gamma(\text{net}(t)) = \gamma\left(\sum_i w_i(t)\,\text{in}_i(t) + \theta(t)\right) \qquad (2.2)$$

Generally, some type of threshold function is used. Such examples are the sign function (sgn), the linear or semilinear function, the ramp function, or the sigmoid function. These functions are presented in Figure 2.3.

The sign function is given by

$$\gamma(s) = \text{sgn}(s) = \begin{cases} +1 & s > 0 \\ -1 & s \leq 0 \end{cases} \qquad (2.3)$$

However, the classic case is the use of the family of sigmoid functions that belong to the class

$$\Gamma = \{\gamma = \gamma(s, k, T, c) \,|\, x, s \in \Re - \{0\}\} \qquad (2.4)$$

and is defined as follows:

$$\gamma(s) = k + \frac{c}{1 + e^{-Ts}} \qquad (2.5)$$

where T is a factor regulating the speed of transition to one of two asymptotic values. This type of function is very important because it provides nonlinearity to the neuron, which is essential in modeling nonlinear phenomena. If in equation (2.5) we set

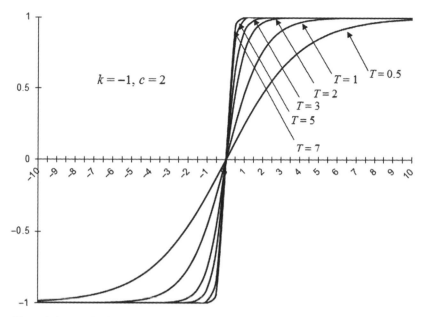

Figure 2.4 *Family of symmetric sigmoid activation functions with values between −1 and 1.*

$k = -1$ and $c = 2$, the symmetric sigmoid activation functions are obtained, which return values between −1 and 1.

As presented in Figure 2.4, if the value of T increases, this family of functions converges to the sign function (2.3). The most commonly used function of this form is obtained for $T = 2$:

$$\gamma(s) = \frac{2}{1 + e^{-2s}} - 1 \qquad (2.6)$$

Similarly, if in equation (2.5) we set $k = 0$ and $c = 1$, we obtain the family of asymmetric sigmoid activation functions which return values between 0 and 1 (Figure 2.5). Usually, for this function the value of $T = 1$ is used:

$$\gamma(s) = \frac{1}{1 + e^{-s}} \qquad (2.7)$$

Connectivity Model

The connectivity model is related to the organization of the connections between units. It determines what the system knows and how it will respond to some random input.

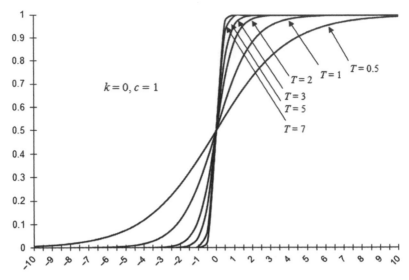

Figure 2.5   *Family of asymmetric sigmoid activation functions with values between 0 and 1.*

Usually, it is assumed that each unit provides an additive contribution to the inputs of the units with which it is connected. Hence, the total (net) input of a unit is simply the weighted sum of inputs from the other units plus a bias term:

$$\text{net} = \sum_i w_i \text{in}_i + \theta \qquad (2.8)$$

Propagation Rule   The propagation rule refers to the way that data flow through a network. There are two basic categories of neural networks: feedforward networks and recurrent networks.

Feedforward Networks   In this case, there is a forward flow of data from the input units to the output units. The processing of the data can be extended to multiple layers of hidden units, but there are no feedback connections. In other words, there are no connections from the outputs of the units to the inputs of the units of the same or previous layers.

Recurrent Network   These networks include feedback connections. Unlike in feedforward networks, in this case the dynamic properties of the network are significant.

Learning Rule   When a set of inputs is inserted in the neural network, it should return the desired set of outputs. To do so, appropriate values of the weights must be selected. One method is to give values to the weights relying on existing (a priori) knowledge of the problem. However, in most cases this knowledge is not available. Another method is to "train" the network by presenting "training examples" and allow the network to change the weights of the connections according to some learning rule. The ways in which learning is conducted generally falls into two broad categories: supervised and unsupervised learning.

In *supervised* or *associative learning*, the network is trained by providing training input and their corresponding output examples. The network gradually learns the underlying relationship between the inputs and the output. In *unsupervised learning* or *self-organizing*, the output units are trained to respond in some complexes of input examples and to discover some of their prominent statistical properties. Unlike supervised learning, there are no a priori categories in which the patterns can be classified, but the network should develop its own representation for the input stimuli. Both types of learning result in an adjustment of the weights of the connections between the units, according to some learning rule.

PERCEPTRON

Suppose that a feedforward neural network has no hidden layers. Hence, the output units are connected directly with the input units. Such a network is shown in Figure 2.6. More precisely, the network shown has only one output unit, two input units, and a bias term, θ.

Furthermore, we assume a training sample that consists of the input vector $\mathbf{x} = (x_1, x_2)$ and the corresponding desired output y. In classification problems y usually takes values of -1 and $+1$. The *perceptron learning algorithm* that updates the weights is the following:

1. Initialize the weights to random values.
2. Select an input vector from the training sample, present it to the network, and compute the output of the network, out, for the input vector specified and the values of the weights.
3. If out $\neq y$ (i.e., the response of the perceptron is incorrect), modify all connections w_i by adding the changes $\Delta w_i = y x_i$.
4. Go back to step 1.

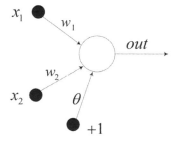

Figure 2.6   *Feedforward neural network without hidden layers with one output and two inputs.*

When the response of the network is correct, the weights are not updated. The weight of the bias term should also be updated. The bias may be seen as the connection w_0 between the output of the network and an entry input that always has the value of 1. Hence, based on the perceptron algorithm, the change of bias is zero if the network response is correct and is equal to y otherwise.

However, in problems where the input–output pairs are not linearly separable, they cannot be modeled with the perceptron rule; the use of neural networks with intermediate hidden layers is required.

The Approximation Theorem

For the perceptron learning rule described above, Rosenblatt (1959) formulated and proved the famous theorem of convergence, which states: If there is a set of connection weights $\mathbf{w}^*$ that is able to perform the transformation out $= y$, the perceptron learning rule will converge to a solution (which may or may not be the same as $\mathbf{w}^*$) after a finite number of steps for any initial selection of weights. For a proof of the theorem the reader is directed to Rosenblatt (1959).

THE DELTA RULE

For a network with no hidden units, one output, and a linear activation function, the output of the network is given by

$$\text{out} = \sum_i w_i x_i + \theta \tag{2.9}$$

Then, for a given input vector output from the training sample, we have that

$$\text{out}_p = \sum_i w_i x_{pi} + \theta \tag{2.10}$$

Such a simple network has the ability to represent a linear relationship between the input and output variables. Using a sign function as an activation function in the output unit, we can construct a classifier, such as the Adaline of Widrow and Hoff (1960). Here we focus on the linear relationship, but we will use the network for a function approximation problem.

We assume that we want to train a network to adapt as best as possible to a hyperplane in a training sample consisting of pairs of the form $(\mathbf{x}_p, y_p)$, where $\mathbf{x}_p = (x_1, x_2, \ldots, x_m)$ and $p = 1, \ldots, n$. For each input vector $\mathbf{x}_p$, the network's output differs from the target value by $y_p - \text{out}_p$. The error function that uses the delta rule is based on the squares of these differences. Specifically, the error for the example p is calculated from the relationship

$$E_p = \tfrac{1}{2}(y_p - \text{out}_p)^2 \tag{2.11}$$

The total error that is minimized by the delta rule is given by the relationship

$$E = \sum_{p=1}^{n} E_p = \frac{1}{2} \sum_{p=1}^{n} (y_p - \text{out}_p)^2 \qquad (2.12)$$

The minimum mean error finds the values of the weights that minimize the error function (2.12) with the *method of gradient descent*, also called the *method of steepest descent*. This method is based on the change in weight w_i by $\Delta_p w_i$, which is proportional to the negative value of the derivative of the error E_p which has been calculated for the training pattern p with respect to the weight w_i, namely:

$$\Delta_p w_i = -\eta \frac{\partial E_p}{\partial w_i} \qquad (2.13)$$

where η is a constant called the *learning rate*. Using the chain rule, the derivative $\partial E_p / \partial w_i$ can be written as

$$\frac{\partial E_p}{\partial w_i} = \frac{\partial E_p}{\partial \text{out}_p} \frac{\partial \text{out}_p}{\partial w_i} \qquad (2.14)$$

The first derivative is easily calculated from (2.11) as follows:

$$\frac{\partial E_p}{\partial \text{out}_p} = -(y_p - \text{out}_p) = -\delta_p \qquad (2.15)$$

where δ_p is the difference between the output of the network and the target value y for pattern p. Because of the linear output unit (2.10), the partial derivative of the output of the network with respect to the weight w_i is

$$\frac{\partial \text{out}_p}{\partial w_i} = x_{pi} \qquad (2.16)$$

Placing equations (2.15) and (2.15) into (2.14), we have that

$$-\frac{\partial E_p}{\partial w_i} = \delta_p x_{pi} \qquad (2.17)$$

Substituting into equation (2.13), we finally get

$$\Delta_p w_i = -\eta \delta_p x_{pi} \qquad (2.18)$$

which express the delta rule. Combining equation (2.17) with the equation

$$\frac{\partial E}{\partial w_i} = \sum_p \frac{\partial E_p}{\partial w_i} \qquad (2.19)$$

we conclude that the change in the weight w_i after a complete cycle of presentation of all the training patterns of the training sample is similar to that of the derivative, and thus the delta rule implements a gradient descent in the space $E - w$. This is true only if the weights change only at the end of the cycle. If the weights change after the presentation of each training example, the method deviates slightly from the true gradient descent. If the learning rate is small enough, this deviation would be negligible and the delta rule would be a very good approximation of the gradient descent to the sum of squared errors. More precisely, if the learning rate is small enough, the delta rule will find a set of weights that minimize the error function.

Although this algorithm is better than the one applied to perceptrons, it cannot be applied to networks that have hidden layers. For each neuron, its output must be known exactly, which is not possible when there are hidden layers.

BACKPROPAGATION NEURAL NETWORKS

As we saw earlier, neural networks without hidden layers are characterized by severe limitations, as the set of problems which they solve is very limited. Minsky and Papert (1969) showed that neural networks with hidden layers can overcome many of these limitations; however, a solution to the problem of adjusting the weights of the connections between the input units and the hidden units was not given. One solution to this problem was given by Rumelhart et al., (1986a).

The idea behind this method is that errors in the hidden units of the hidden layer are determined by the backpropagation of errors in the output units. This is why this training algorithm is called the *backpropagation learning rule*. The backpropagation algorithm can be seen as a generalization of the delta rule for the case of neural networks with hidden layers and nonlinear activation functions.

Multilayer Feedforward Networks

The processing units of feedforward networks are organized in layers, which is why they are called *multilayer networks*. The first of these layers, the *input layer*, is used for data entry. The processing units of this layer do not perform any computations (they do not have any input weights or activation functions).

Next are one or more intermediate hidden layers, then an *output layer*. The units of each layer receive their inputs from the units of the layer immediately below (behind) and send their outputs to the units of the layer lying directly above (front).

Information flow

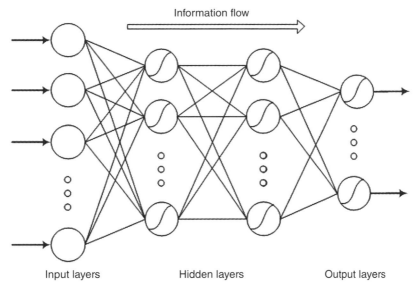

Input layers   Hidden layers   Output layers

Figure 2.7   *Fully connected feedforward neural network with two hidden layers.*

The network can be connected either fully or partially. In the former case, the processing units of different layers are associated with all the units of the next layer, whereas in the Latter case the connection is with only some of them. There are no feedback connections (i.e., connections that send the output of a unit in the same or a previous layer). Also, there are no connections between units of the same layer. The flow of information is done through the connections and the processing units in one direction only: from the input layer to the output layer. In Figure 2.7, a schematic representation of a fully connected feedforward neural network with two hidden layers is presented.

Although the backpropagation learning rule may be applied to feedforward networks with any number of hidden layers, a series of studies (Funahashi, 1989; Hartman et al., 1990; Hornik et al., 1989) have proved that a single hidden layer is sufficient for the neural network to approximate any function to any random degree of accuracy, with the condition that the activation functions of the network are nonlinear. This is known as the universal approximation theorem. In most cases, neural networks with a single hidden layer and sigmoidal transfer function are used.

THE GENERALIZED DELTA RULE

The backpropagation method is the most widely used method for training multilayer neural networks. The basic idea behind this training algorithm is to determine the percentage of the total error for the weight of each connection. Hence, it is possible to calculate the correction to the weight of each connection separately, which is quite complex for the hidden layers since their outputs affect many connections simultaneously.

In backpropagation, initially the error of the output unit is estimated in the same way as for the delta rule. This error is used to calculate the errors in the last hidden layer. Then the process is repeated recursively toward the input layer. Based on the propagation of the error backward, it is possible to estimate the contribution of each connection to the total error. Then the errors estimated for each connection of the layer are used to alter the weights of the connections in a manner similar to that of the delta rule. The process is therefore based on a generalization of the delta rule, which is why it is called the generalized delta rule. The procedure is repeated until the value of the total error reaches a predefined value.

As discussed earlier, the delta rule performs a gradient descent on the sum of squared errors for linear transfer functions. In the case of no hidden units, error surface has a convex shape with a unique minimum. It is therefore it is guaranteed that the gradient descent will eventually find the best set of weights. However, when the neural network has hidden units, the error surface does not have a unique minimum. As a result, the alogrithm might reach and be trapped into a local minimum.

In the delta rule algorithm, first a set of training patterns are presented to the network. The training patterns consist of the input and output vectors. First, the network uses the input vector to calculate its own output vector and then it compares it against the desired output vector (the target). If there is no difference, the learning procedure stops. Otherwise, the weights of the network connections are modified to reduce the difference. If we do not have any hidden layers, the network connections change according to the delta rule. More precisely, the change in the weight of the connection between the input and output units is given by (2.18).

In the remainder of the section we present the generalized delta rule for multilayer feedforward neural networks with nonlinear activation functions. The net input of an output unit u_j in Figure 2.8 for the output–input vector p is given by

$$\text{net}_{pj} = \sum_i w_{ij}\text{out}_{pi} \tag{2.20}$$

To calculate the output of the same unit, an upward, continuous, and differentiable function is used:

$$\text{out}_{pj} = \gamma(\text{net}_{pj}) \tag{2.21}$$

The usual function that is used is the sigmoid. In the generalized delta rule the changes in the weights are given by

$$\Delta_p w_i = -\eta \frac{\partial E_p}{\partial w_{ij}} \tag{2.22}$$

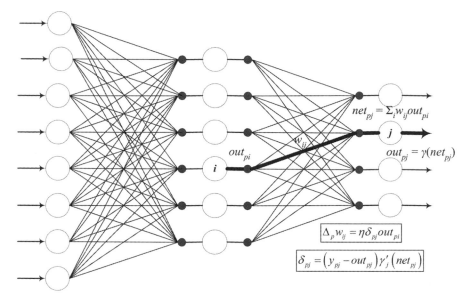

Figure 2.8 *Estimation of the delta in a feedforward neural network for the connections to the output units of the network.*

where E is the same measurement of the square errors used earlier. Using the chain rule, the derivative $\partial E / \partial w_{ij}$ can be written as

$$\frac{\partial E}{\partial w_{ij}} = \frac{\partial E}{\partial \text{net}_{pj}} \frac{\partial \text{net}_{pj}}{\partial w_{ij}} \qquad (2.23)$$

From equation (2.20) we see that the partial derivative of the total input to unit j based on the weight of one of the connections w_{ij} is

$$\frac{\partial \text{net}_{pj}}{\partial w_{ij}} = \frac{\partial}{\partial w_{ij}} \left(\sum_k w_{kj} \text{out}_{pk} \right) = \text{out}_{pi} \qquad (2.24)$$

Next, we define

$$\delta_{pj} = -\frac{\partial E_p}{\partial \text{net}_{pj}} \qquad (2.25)$$

This definition is consistent with the definition of the delta given by equation (2.18) since for linear transfer functions we have that $\text{net}_{pj} = \text{out}_{pj}$.

From relationships (2.23) to (2.25) we have that

$$-\frac{\partial E_p}{\partial w_{pj}} = \delta_{pj} \text{out}_{pj} \tag{2.26}$$

and by replacing (2.26) with (2.22) we observe that the changes in the weights of the connections can be computed by

$$\Delta_p w_{pj} = \eta \delta_{pj} \text{out}_{pj} \tag{2.27}$$

From equation (2.27) we observe that the change in weight w_{ij} of the connection between the units u_i and u_j (the unit u_j is located in the next layer of unit u_i) is a product of the learning rate and the delta of unit u_j and the output unit u_i. The learning rate is a constant chosen by us, and the output of unit i is easily calculated from equation (2.21). Hence, we are interested in estimation of the delta of each unit. Applying the chain rule in (2.25), we have that

$$\delta_{pj} = -\frac{\partial E_p}{\partial \text{net}_{pj}} = -\frac{\partial E_p}{\partial \text{out}_{pj}} \frac{\partial \text{out}_{pj}}{\partial \text{net}_{pj}} \tag{2.28}$$

From (2.21) we see that the second term of the product in (2.28) is just the derivative of the transfer function of the unit u_j:

$$\frac{\partial \text{out}_{pj}}{\partial \text{net}_{pj}} = \gamma'(\text{net}_{pj}) \tag{2.29}$$

To estimate the first term of (2.28) we have to distinguish between two cases. First we consider the case where the unit u_j is located in the output layer, as in Figure 2.8. From the definition of the error criterion we have that

$$\frac{\partial E_p}{\partial \text{out}_{pj}} = -(y_{pj} - \text{out}_{pj}) \tag{2.30}$$

Therefore, substituting relations (2.30) and (2.29) into (2.28) we find that the delta of the output units of the network is calculated as follows:

$$\delta_{pj} = (y_{pj} - \text{out}_{pj}) \gamma'(\text{net}_{pj}) \tag{2.31}$$

Next, we focus in the case where the unit u_j is located in a hidden layer, as in Figure 2.9. In this case the contribution of the output of the unit to the error E_p cannot be calculated directly. However, measurement of the error is a function of the net input to the units of the output layer:

$$E_p = E_p(\text{net}_{p1}, \text{net}_{p2}, \ldots, \text{net}_{pk}, \ldots) \tag{2.32}$$

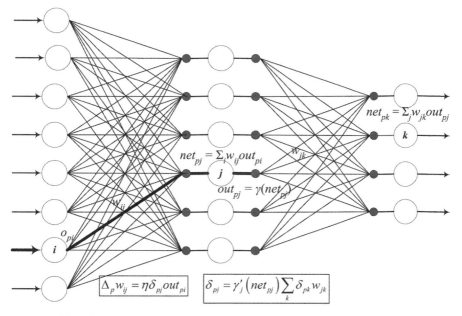

Figure 2.9 *Estimation of the delta of a feedforward network in the hidden layer.*

Applying the chain rule, we have

$$\frac{\partial E_p}{\partial \text{out}_p} = \sum_k \frac{\partial E_p}{\partial \text{net}_{pk}} \frac{\partial \text{net}_{pk}}{\partial \text{out}_p}$$

$$= \sum_k \frac{\partial E_p}{\partial \text{net}_{pk}} \frac{\partial}{\partial \text{out}_p} \left(\sum_j w_{jk} \text{out}_{pj} \right)$$

$$= \sum_k \frac{\partial E_p}{\partial \text{net}_{pk}} w_{jk}$$

$$= - \sum_k \delta_{pk} w_{jk} \tag{2.33}$$

By replacing equations (2.29) and (2.33) with (2.28), we get

$$\delta_{pj} = \gamma'_j(\text{net}_{pj}) \sum_k \delta_{pk} w_{jk} \tag{2.34}$$

BACKPROPAGATION IN PRACTICE

The generalized delta rule is applied in two steps. During the first step, the input vector **x** is introduced to the network and is propagated from layer to layer toward the output layer and the network's outputs are computed. The outputs are compared with the target values desired, creating an error signal from each output of the network. At the second step, the error signal propagates backward through the network (i.e., from the output to the input layer). At the same time, the appropriate changes in the weights of the network's connections are estimated.

These two steps are summarized as follows:

- For a given input vector and initial values of the weights of the network connections, the net inputs and outputs of each unit are estimated, moving from the input layer to the output layer. Finally, the approximation error is estimated.
- Then, the derivatives of the transfer functions and the delta of the output units are estimated.

The most commonly used activation function is the asymmetrical sigmoid with asymptotic values 0 and 1:

$$\gamma(\text{net}_{pj}) = \frac{1}{1 + e^{-\text{net}_{pj}}} \tag{2.35}$$

The derivative of this transfer function with respect to net_{pj} is given by

$$
\begin{aligned}
\gamma'(\text{net}_{pj}) &= \frac{d}{d\,\text{net}_{pj}} \frac{1}{1 + e^{-\text{net}_{pj}}} \\
&= \frac{e^{-\text{net}_{pj}}}{(1 + e^{-\text{net}_{pj}})^2} \\
&= \frac{1}{(1 + e^{-\text{net}_{pj}})} \frac{e^{-\text{net}_{pj}}}{(1 + e^{-\text{net}_{pj}})} \\
&= \gamma(\text{net}_{pj})[1 - \gamma(\text{net}_{pj})]
\end{aligned} \tag{2.36}
$$

As we see the derivative of the activation function is expressed as a function of itself. Substituting the relationship above to the delta, we have that

$$\delta_{pj} = (y_{pj} - \text{out}_{pj})\gamma_j(\text{net}_{pj})[1 - \gamma_j(\text{net}_{pj})] \tag{2.37}$$

- Next, the changes in the weights of connections of the output layer are computed using the generalized delta rule.
- Next, the derivatives of the transfer functions and the delta of the units of the previous hidden layer are computed.

$$\delta_{pj} = \gamma_j(\text{net}_{pj})[1 - \gamma_j(\text{net}_{pj})] \sum_k \delta_{pk} w_{jk} \tag{2.38}$$

- Then the changes of weights of the connections of the previous hidden layer are computed using the generalized delta rule.
- The previous two steps are repeated for all hidden layers moving from the output layer to the input layer.

TRAINING WITH BACKPROPAGATION

The backpropagation algorithm implements a search for the total minimum error of the function $E(\mathbf{w})$, which has as parameters the values of the weights. In each step the weights are corrected by choosing a change that seems to reduce the error locally with the aim of minimizing the error $E(\mathbf{w})$. Given the current weight vector $\mathbf{w}_c$, for each iteration a direction $\mathbf{u}_c$ is calculated and then the weight vector $\mathbf{w}_c$ is updated using the following learning rule:

$$\mathbf{w}_+ = \mathbf{w}_c - \eta \mathbf{u}_c \qquad (2.39)$$

where $\mathbf{w}_+$ is the new weight vector and η is the learning rate. Specifically, the backpropagation algorithm calculates the slope $\nabla E(\mathbf{w}_c) = \partial E / \partial \mathbf{w}_c$ and then performs a minimization step toward the direction $\mathbf{u}_c = \nabla E(\mathbf{w}_c)$.

Note that the gradient operator ∇ is meaningless by itself. On the other hand, the gradient vector $\nabla E(\mathbf{w}_c)$ points in the direction of the maximum growth rate of the error function at point $\mathbf{w}_c$ and is equal to this growth rate. Hence, the learning rule is

$$\mathbf{w}_+ = \mathbf{w}_c - \eta \nabla E(\mathbf{w}_c) \qquad (2.40)$$

As presented in (2.39), the product between $\mathbf{u}_c$ and the learning rate is subtracted from the current weight vector. When $\nabla E(\mathbf{w}_+)$ is vertical to $\mathbf{u}_c$, the backpropagation algorithm has found a minimum $\mathbf{w}_{min}$.

In Figure 2.10 a simplistic schematic representation of the gradient descent is shown. In reality, the surface $(E - \mathbf{w})$ cannot be represented graphically when there are two more weights: in other words, for any useful neural network topology.

However, the greater the value of the learning rate η, the greater the risk that the stepwise gradient descent starts oscillating, as presented in Figure 2.11. In this case it is impossible to find the global minimum as the algorithm oscillates between two points (e.g., in points 3 and 4 in Figure 2.11).

On the other hand, very small values of the learning rate drastically reduce the chances that the algorithm will start oscillating. However, the algorithm becomes significantly slower. A schematic representation of the behavior of backpropagation algorithm with a relatively small learning rate is presented in Figure 2.12.

To avoid oscillations and at the same time to speed up the learning process, a momentum term is added to equation (2.27):

$$\Delta_p w_{ij}(t+1) = \eta \delta_{pj} \text{out}_{pi} + m \Delta_p w_{ij}(t) \qquad (2.41)$$

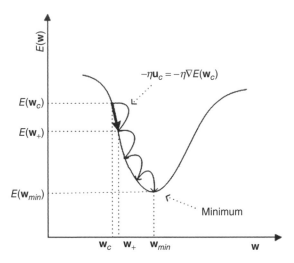

Figure 2.10   *Simplistic representation of the minimum error search through stepwise gradient descent in the weights–error surface ($E − \mathbf{w}$) that is implemented by the backpropagation algorithm.*

where m is the momentum term and t refers to the iteration number. From equation (2.41) we observe that the change in the weight estimated in the previous presentation of the training patterns is multiplied by a constant (the momentum) and then is added to the current change of the weight. The momentum determines the effect of the previous change to the next. The addition of the momentum term allows us to use relatively small values for the learning rate without increasing the learning time significantly.

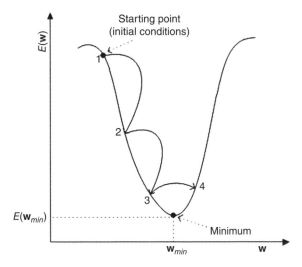

Figure 2.11   *Stepwise descent to the surface ($E − \mathbf{w}$) with a very large learning rate. The backpropagation algorithm starts swinging between points 3 and 4, so it is not possible to find the point that minimizes the error function.*

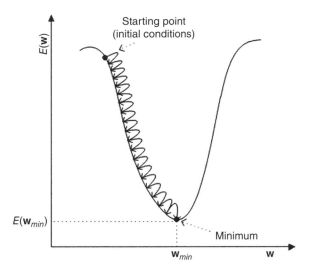

Figure 2.12 *Stepwise descent to the surface ($E - \mathbf{w}$) with a very small learning rate, which results in long training times (more iterations).*

Since in the change of weight a proportion of the previous variation is added, when the algorithm enters a region of the surface ($E - \mathbf{w}$) with a high gradient, it begins to perform a continuously accelerated descent until it reaches the minimum (Figure 2.13). Hence, the relationship (2.40) with the momentum term becomes

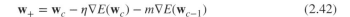

$$\mathbf{w}_+ = \mathbf{w}_c - \eta \nabla E(\mathbf{w}_c) - m \nabla E(\mathbf{w}_{c-1}) \qquad (2.42)$$

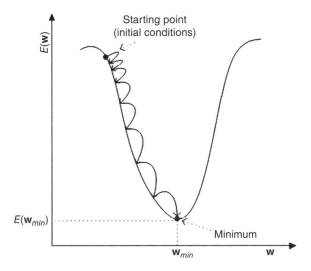

Figure 2.13 *Stepwise descent to the surface ($E - \mathbf{w}$) with a relatively small learning rate and momentum term, which accelerates the search for the minimum of the error function.*

In general, the purpose of the addition of the momentum term is to accelerate the network's training process without a disproportionate increase the probability that the algorithm will be trapped in oscillations between two nonoptimal solutions.

Despite the obvious potential of the backpropagation algorithm, under some circumstances various problems can occur. It is possible, for example, for the algorithm to reach a static state known as *network paralysis*, due to the saturation of the activation function. Additional problems arise due to the existence of *local minima* in the weights–error surface and due to the nonuniqueness of the solutions. Finally, a common problem that can occur is long training times, as a result of the suboptimal choice of the learning rate and the momentum values. This can be treated by extending the backpropagation algorithm and using a variable learning rate and momentum. We examine these problems and how to tackle them next.

Network Paralysis

One problem that may possibly arise in the training of the network is for the backpropagation algorithm to reach a stationary state. Network paralysis may occur due to the values that the network's weights may reach. If the values of the weights become too high, this will also lead to very high values of the net input of some hidden units. Furthermore, if a sigmoid activation function is used, the output of the unit will be very close to 1. In this case the delta estimated from relations (2.37) and (2.38) will be very close to zero, as $\gamma_j(\text{net}_{pj})[1 - \gamma_j(\text{net}_{pj})] \approx 0$. Hence, the changes of the weights calculated from relation (2.27) would be practically zero. So, in reality, the learning process stops.

The same problem would arise in the event that some weights had very low values. Then the net input of some units will be very small, and as a result the output of the unit will be very close to zero. Since $\gamma_j(\text{net}_{pj})[1 - \gamma_j(\text{net}_{pj})] \approx 0$, the delta would also be negligible, as would the changes in weights.

Wang et al., (2004) proposed a solution to the problem of network paralysis. For each training sample they suggested using a different activation function. In addition, the activation functions would be adjusted continuously during training to avoid saturation.

Local Minima

Another problem that arises during the training of a network is associated with the morphology of the weight–error surface $(E - \mathbf{w})$, which for feedforward neural networks with hidden layers is generally very complicated. A simplistic two-dimensional representation is shown in Figure 2.14, where we see that there is more than one minimum (points B, C, and D). The backpropagation algorithm can reach one of these solutions, depending on the starting point (points A and E).

The way in which the stepwise gradient descent is implemented does not provide any guarantee that the solution found will be the overall optimal, that is, that it will correspond to the lower overall error level (point C in Figure 2.14). This solution is known as the *global minimum*; other solutions are called *local minima*. The inherent

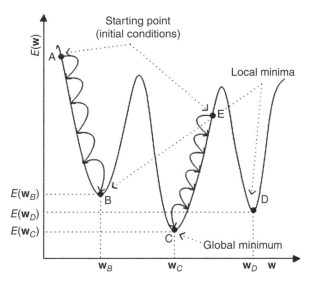

Figure 2.14 *Surface weight–error ($E - \mathbf{w}$) with multiple minima. The backpropagation algorithm can be "trapped" in a local minimum (point B) which does not correspond to the total minimum error (point C), depending on the starting point of the gradient descent.*

weakness of the gradient descent to find the global minimum is due to the selection of the architecture (topology) of the neural network in relation to the complexity of the problem. When the complexity of the network is greater than required for a relevant application and for the size of the training sample, the estimated model will suffer from a high level of variance. This means that for the same network architecture, repeated sampling from the training sample, then fitting a model to the new sample, will result in quite different solutions: in other words, to different weight vectors. On the other hand, when the network architecture is simpler than necessary—that is, the number of network parameters is relatively small—the fitted model runs the risk of being underparameterized or *biased*.

A simple method that is used to increase the chances of finding the global minimum is called *weight jogging*. In this method, when an initial solution is found, a small and random change is added to the weight vector and the training process continues until the algorithm converges to a new solution. A simple example of weight jogging is shown in Figure 2.15. The backpropagation algorithm has converged initially at point B, which corresponds to a local minimum. In the weight vector of the solution $\mathbf{w}_B$, a small random change is added. The modified vector $\mathbf{w}_C$ corresponds, as expected, to a greater level of error, but is located such that if network training continues, it will lead to the global minimum D.

Of course, there is no guarantee that this process will lead to a better solution, as it is simply an improvement in the local search carried out by the backpropagation algorithm. Global search algorithms, as the simulated annealing, deal much better with the problem of local minima.

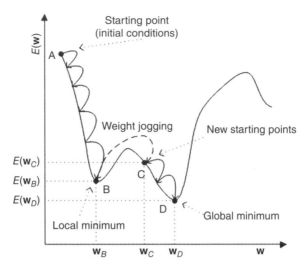

Figure 2.15   *Small and random changes in the original solution (weight jogging) increase the chances of finding the global minimum.*

Nonunique Solutions

Whether the solution corresponds to a local or a global minimum, it is sometimes not unique. This means that there are multiple combinations of weights and/or processing units (i.e., more than one network topology) that correspond to the same value of the error function (Chen and Hecht-Nielsen, 1991; Chen et al., 1993).

In general, we may have different local minima, derived from different weights that correspond to the same error level. This case is presented in Figure 2.16 in solutions Z and G, where $E_2 = E(\mathbf{w}_Z) = E(\mathbf{w}_G)$. Different initial conditions lead to different solutions but are equivalent in the sense of minimizing the error function.

Another case is to have many similar solutions corresponding to the same level of error and to create a flat minimum. This is the plateau that is shown in Figure 2.16, where the adjacent weight vectors $\mathbf{w}_B$, $\mathbf{w}_C$, and $\mathbf{w}_D$ correspond to the same error, $E_1 = E(\mathbf{w}_B) = E(\mathbf{w}_C) = E(\mathbf{w}_D)$. The existence of such plateaus is a characteristic of the overparameterized networks, where the complexity of their architecture exceeds the requirements of the particular application. This problem is addressed by such prunning methods as the ICE (Zapranis and Haramis, 2001), OBS (Hassibi and Stork, 1993), and OBD (LeCun et al., 1989), algorithms.

CONFIGURATION REFERENCE

Neural networks are nonlinear estimators; hence, there are no a priori assumptions regarding the structure of the model. As presented earlier, the stepwise algorithms used for training of the neural networks are known as learning algorithms, due to their iterative nature. White (1989) has demonstrated that the learning algorithms

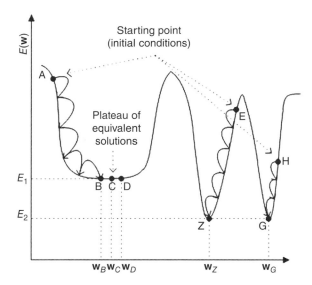

Figure 2.16 *Weight–error surface (E − w) with a flat minimum. Various vector weights w_B, w_C, and w_D of the solutions B, C, and D, corresponding to the same level of error, E_1. Also, the local minima G and I correspond to the same level error, E_2.*

of neural networks can be formulated as a nonlinear regression problem. We have also demonstrated that the main strength of the neural networks lies in the universal approximation property. Furthermore, 1 hidden layer is sufficient for a backpropagation neural network to approximate to any degree any function and its derivatives.

The usual structure of a backpropagation neural network that is used in the majority of the applications is presented in Figure 2.17. The network has only 1 hidden layer with λ hidden units. The input layer consists of m units, without any activation function. The input signal of these units is identical to the output signal. All other units of the neural network use the asymmetric sigmoidal activation function with values in (0, 1). In addition, in the input and output layers, a bias unit is added. These units do not accept any input, and their output has a constant value of 1. The desired output of the network can be continuous or discrete.

The number m of the input units is determined by the relevant problem. The term *architecture* or *topology* of a network, A_λ, refers to the topological arrangement of the network's connections. We define a class of neural networks S_λ:

$$S_\lambda \equiv \{g_\lambda(\mathbf{x}; \mathbf{w}), \mathbf{x} \in \mathfrak{R}^m, \mathbf{w} \in \mathbf{W}, \mathbf{W} \subseteq \mathfrak{R}^p\} \tag{2.43}$$

where $g_\lambda(\mathbf{x}; \mathbf{w})$ is a nonlinear function, $\mathbf{w} = (w_1, w_2, \ldots, w_p)$ is the vector of parameters (i.e., the weights of network connections), and p is the number of parameters determined by the topology A_λ.

The class of neural networks S_λ is a set of neural networks that have the same architecture and whose members are differentiated and simultaneously defined entirely

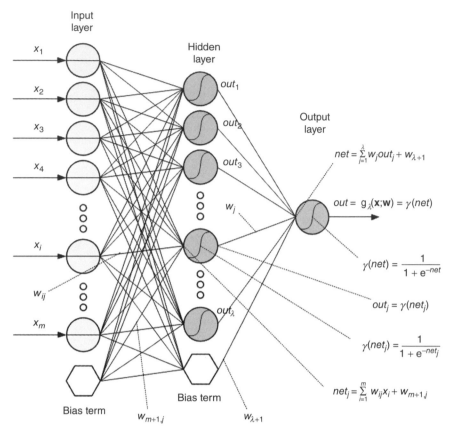

Figure 2.17 *Backpropagation neural network with 1 hidden layer and 1 output unit.*

by the weight vector $\mathbf{w}$. In the case of a neural network with a 1-hidden-layer architecture (Figure 2.17), the number of hidden units λ defines the different classes S_λ since it uniquely determines the dimension of the parameter vector p, where $p = (m + 2)\,\lambda + 1$.

For this type of network, given an input vector $\mathbf{x}$ and a weight vector $\mathbf{w}$, the output of the network $g_\lambda(\mathbf{x};\mathbf{w})$ is

$$g_\lambda(\mathbf{x};\mathbf{w}) = \gamma\left(\sum_{j=1}^{\lambda} w_j \gamma\left(\sum_{i=1}^{m} w_{ij}x_i + w_{m+1,j}\right) + w_{\lambda+1}\right) \qquad (2.44)$$

where w_{ij} is the weight of the connection between the input i and the hidden unit j, $w_{m+1,j}$ is the weight of the connection between the term bias in the input layer $m + 1$ and the hidden unit j, w_j is the weight of the connection between the hidden unit j and the output unit, and $w_{\lambda+1}$ is the weight of the connection between the bias term of the hidden layer $\lambda + 1$ unit and the output.

CONCLUSIONS

Artificial neural networks (or simply, neural networks) derive their origins from biological neural networks. They are systems of parallel and distributed processing that simulate the basic operating principles of the biological brain.

Feedforward neural networks with at least 1 hidden layer and nonlinear activation functions have the property of universal approximation. In other words, they can approximate any function. The stepwise algorithm that is used for their training is based on the generalization of the delta rule that is used in perceptron networks. Updating the weights of the network is done by propagating the error backward, so this method is known as the backpropagation algorithm and the networks as backpropagation neural networks. The backpropagation algorithm, despite its disadvantages, is the most commonly used training algorithm, as it is simpler and significantly less computationally expensive.

From a statistical perspective, neural networks can be formulated as a nonparametric nonlinear regression model. In this book we treat neural networks as the eminent expression of nonparametric regression, which constitutes a very powerful approach, especially for financial applications. The main characteristic of neural networks is their ability to approximate any nonlinear functions without making a priori assumptions about the nature of the process that created the available observations. This is particularly useful in financial applications, where there are many hypotheses but little is actually known about the nature of the processes that determine the prices of assets. Nevertheless, knowledge of the relative theory of neural learning and the basic models of neural networks are essential for their effective implementation, especially in complex applications such as the ones encountered in finance.

REFERENCES

Chen, A. M., and Hecht-Nielsen, R. (1991). "On the geometry of feedforward neural network weight spaces." *Second International Conference on Artificial Neural Networks*, 1–4.

Chen, A. M., Lu, H.-M., and Hecht-Nielsen, R. (1993). "On the geometry of feedforward neural network error surfaces." *Neural Computation*, 5(6), 910–927.

Funahashi, K.-I. (1989). "On the approximate realization of continuous mappings by neural networks." *Neural Networks*, 2(3), 183–192.

Hartman, E. J., Keeler, J. D., and Kowalski, J. M. (1990). "Layered neural networks with Gaussian hidden units as universal approximations." *Neural Computation*, 2(2), 210–215.

Hassibi, B., and Stork, D. G. (1993). "Second order derivatives for network pruning: optimal brain surgeon." *Advances in Neural Information Processing Systems*, 164–164.

Hornik, K., Stinchcombe, M., and White, H. (1989). "Multilayer feedforward networks are universal approximators." *Neural Networks*, 2(5), 359–366.

LeCun, Y., Denker, J. S., Solla, S. A., Howard, R. E., and Jackel, L. D. (1989). "Optimal brain damage." *NIPS*, 598–605.

Minsky, M., and Papert, S. (1969). *Perceptron: an Introduction to Computational Geometry*, expanded edition. MIT Press, Cambridge, MA, pp. 19, 88.

Rosenblatt, F. (1959). *Principles of Neurodynamics.* Spartan Books, New York.

Rumelhart, D. E., Hintont, G. E., and Williams, R. J. (1986a). "Learning representations by back-propagating errors." *Nature*, 323(6088), 533–536.

Rumelhart, D. E., McClelland, J. L., and University of California–San Diego PDP Research Group. (1986b). *Parallel Distributed Processing: Explorations in the Microstructure of Cognition*, MIT Press, Cambridge, MA.

Wang, X., Tang, Z., Tamura, H., Ishii, M., and Sun, W. (2004). "An improved backpropagation algorithm to avoid the local minima problem." *Neurocomputing*, 56, 455–460.

White, H. (1989). "Learning in artificial neural networks: a statistical perspective." *Neural Computation*, 1, 425–464.

Widrow, B., and Hoff, M. E. (1960). "Adaptive switching circuits." *IRE WESCON Convention Report*, 96–104.

Zapranis, A. D., and Haramis, G. (2001). "An algorithm for controlling the complexity of neural learning: the irrelevant connection elimination scheme." *Fifth Hellenic European Research on Computer Mathematics and Its Applications*, Athens, Greece.

3

Wavelet Neural Networks

In the literature, various versions of wavelet networks have been proposed. A wavelet network usually has the form of a three-layer network. The lower layer represents the input layer, the middle layer is the hidden layer, and the upper layer is the output layer. The way that the three layers are connected and interact defines the structure of the network. In the input layer, the explanatory variables are inserted in the model and transformed to wavelets. The hidden layer consists of wavelons, or hidden units. Finally, all the wavelons are combined to produce the output of the network, $\hat{y}_p$, at the output layer, which is an approximation of the target value, y_p.

In this chapter we present and discuss analytically the structure of the wavelet network proposed. More precisely, in this book, a multidimensional WN with a linear connection between the wavelons and the output is implemented. In addition, there is a direct connection from the input layer to the output layer that will help the network to perform well in linear applications. In other words, a wavelet network with zero hidden units is reduced to a linear model.

Furthermore, the initialization phase, the training phase, and the stopping conditions are discussed. A wavelet is a waveform of effectively limited duration that has an average value of zero and localized properties. Hence, in wavelet networks, selecting initial values of the dilation and translation parameters randomly may not be suitable, since random initialization may lead to wavelons with a value of zero. Four methods for the initialization of the parameters of a wavelet network are presented and evaluated. The simplest is the *heuristic method*. More sophisticated methods,

Wavelet Neural Networks: With Applications in Financial Engineering, Chaos, and Classification,
First Edition. Antonios K. Alexandridis and Achilleas D. Zapranis.
© 2014 John Wiley & Sons, Inc. Published 2014 by John Wiley & Sons, Inc.

such as residual-based selection, selection by orthogonalization, and backward elimination, can be used for efficient initialization. The parameters of the wavelet network are further optimized during the training phase. In the training phase the parameters are changed to minimize an error function between the target values and the wavelet output. This is done iteratively until the one of the stopping conditions is met.

WAVELET NEURAL NETWORKS FOR MULTIVARIATE PROCESS MODELING

Structure of a Wavelet Neural Network

In this section the structure of a wavelet network is presented and discussed. A wavelet network usually has the form of a three-layer network. The lower layer represents the input layer, the middle layer is the hidden layer, and the upper layer is the output layer. In the input layer the explanatory variables are introduced to the wavelet network. The hidden layer consists of hidden units (HUs). The hidden units, also referred to as wavelons, are similar to neurons in the classical sigmoid neural networks. In the hidden layer the input variables are transformed to a dilated and translated version of the mother wavelet. Finally, in the output layer, the approximation of the target values is estimated.

Various structures of a wavelet network have been proposed. The idea of a wavelet network is to adapt the wavelet basis to the training data. Hence, the wavelet estimator is expected to be more efficient than a sigmoid neural network (Zhang, 1993). An adaptive wavelet network was used by Billings and Wei (2005), Kadambe and Srinivasan (2006), Mellit et al. (2006), and Xu and Ho (1999). Chen et al. (2006) proposed a local linear wavelet network. The difference is that the weights of the connections between the hidden layer and the output layer are replaced by a local linear model. Fang and Chow (2006) and Jiao et al. (2001) proposed a multiwavelet neural network. In this structure, the activation function is a linear combination of wavelet bases instead of the wavelet function. During the training phase, the weights of all wavelets are updated. The multiwavelet neural network is also enhanced by the discrete wavelet transform. Their results indicate that the model proposed increases the approximation capability of the network. Khayamian et al. (2005) introduced a principal component–wavelet neural network. In this context, first principal component analysis has been applied to the training data to reduce the dimensionality. Then a wavelet network was used for function approximation. Zhao et al. (1998) used a multidimensional wavelet-basis function network. More precisely, Zhao et al. (1998) used a multidimensional wavelet function as the activation function in the hidden layer. Then the sigmoid function was used as an activation function in the output layer. Becerikli (2004) proposes a network with unconstrained connectivity and with dynamic elements (lag dynamics) in its wavelet-processing units, called a dynamic wavelet network.

In this study we implement a multidimensional wavelet network with a linear connection between the wavelons and the output. Moreover, for the model to perform

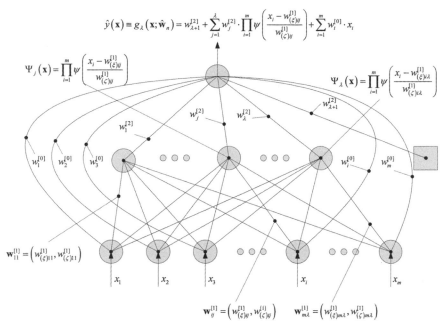

Figure 3.1 *Feedforward wavelet neural network.*

well in the presence of linearity, we use direct connections from the input layer to the output layer. Hence, a network with zero hidden units is reduced to a linear model.

The structure of a single-hidden-layer feedforward wavelet network is given in Figure 3.1. The network output is given by the expression

$$g_\lambda(\mathbf{x}; \mathbf{w}) = \hat{y}(\mathbf{x}) = w_{\lambda+1}^{[2]} + \sum_{j-1}^{\lambda} w_j^{[2]} \cdot \Psi_j(\mathbf{x}) + \sum_{i-1}^{m} w_i^{[0]} \cdot x_i \qquad (3.1)$$

where $\Psi_j(\mathbf{x})$ is a multidimensional wavelet constructed by the product of m scalar wavelets, $\mathbf{x}$ the input vector, m the number of network inputs, λ the number of hidden units, and w a network weight. The multidimensional wavelets are computed as follows:

$$\Psi_j(\mathbf{x}) = \prod_{i=1}^{m} \psi(z_{ij}) \qquad (3.2)$$

where ψ is the mother wavelet and

$$z_{ij} = \frac{x_i - w_{(\xi)ij}^{[1]}}{w_{(\zeta)ij}^{[1]}} \qquad (3.3)$$

In the expression above, $i = 1, \ldots, m, j = 1, \ldots, \lambda + 1$, and the weights w correspond to the translation $(w_{(\xi)ij}^{[1]})$ and the dilation $(w_{(\zeta)ij}^{[1]})$ factors. The complete vector of the network parameters comprises $w = (w_i^{[0]}, w_j^{[2]}, w_{\lambda+1}^{[2]}, w_{(\xi)ij}^{[1]}, w_{(\zeta)ij}^{[1]})$. These parameters are adjusted during the training phase.

In the literature, three mother wavelets are usually suggested: the *Gaussian derivative*, given by

$$\psi(z_{ij}) = z_{ij} e^{(1/2)z_{ij}^2} \tag{3.4}$$

the second derivative of the Gaussian, the *Mexican hat*,

$$\psi(z_{ij}) = \left(1 - z_{ij}^2\right) e^{-(1/2)z_{ij}^2} \tag{3.5}$$

and the *Morlet wavelet*, given by

$$\psi(z_{ij}) = e^{-(1/2)z_{ij}^2} \cos 5z_{ij} \tag{3.6}$$

Selection of the mother wavelet depends on the application and is not limited to the foregoing choices. The activation function can be a wave net (orthogonal wavelets) or a wave frame (continuous wavelets). Following Becerikli et al. (2003), Billings and Wei (2005), and Zhang (1994), we use as a mother wavelet the Mexican hat function, which proved to be useful and to work satisfactorily in various applications.

Initialization of the Parameters of the Wavelet Network

In wavelet networks, in contrast to neural networks that use sigmoid functions, selecting initial values of the dilation and translation parameters randomly may not be suitable (Oussar et al., 1998). A wavelet is a waveform of effectively limited duration that has an average value of zero and localized properties; hence, a random initialization may lead to wavelons with a value of zero. Training algorithms such as gradient descent with random initialization are inefficient (Zhang, 1993), since random initialization affects the speed of training and may lead to a local minimum of the loss function (Postalcioglu and Becerikli, 2007). Also, in sigmoid neural networks, although a minimization of the loss function can be replicated with random initialization, the values of the weights will vary each time (Anders and Korn, 1999).

Utilizing the information that can be extracted by wavelet analysis from the input data set, the initial values of the parameters w of the network can be selected in an efficient way. Efficient initialization will result in fewer iterations in the training phase of the network and training algorithms that will avoid local minimums of the loss function in the training phase. Finally, efficient initialization methods will approximate the same vector of weights that minimize the loss function each time.

Various methods have been proposed for optimized initialization of the wavelet parameters. In the rest of the chapter the following methods are discussed: the heuristic, residual-based selection (RBS), selection by orthogonalization (SSO), and backward elimination (BE).

The Heuristic Initialization Method   The following initialization for the translation and dilation parameters was introduced by Zhang and Benveniste (1992):

$$w^{[1]}_{(\xi)ij} = 0.5(N_i + M_i) \tag{3.7}$$

$$w^{[1]}_{(\zeta)ij} = 0.2(M_i - N_i) \tag{3.8}$$

where M_i and N_i are defined as the maximum and minimum of input x_j:

$$M_i = \max_{p=1,\ldots,n} (x_{ip}) \tag{3.9}$$

$$N_i = \min_{p=1,\ldots,n} (x_{ip}) \tag{3.10}$$

In the framework above, the initialization of the parameters is based on the input domains defined by examples of the training sample (Oussar et al., 1998). In other words, the center of the wavelet j is initialized at the center of the parallelepiped defined by the input domain. These initializations guarantee that the wavelets extend initially over the entire input domain (Oussar and Dreyfus, 2000). This procedure is very simple and the computational burden is almost negligible. Initialization of the direct connections $w^{[0]}_i$ and the weights $w^{[2]}_j$ is less important, and they are initialized in small random values between 0 and 1.

Complex Initialization Methods   The previous heuristic method is simple and its computational cost is almost negligible. However, it is not efficient, as shown on the next section. The heuristic method does not guarantee that the training will find the global minimum. Moreover, this method does not use any information that the wavelet decomposition can provide.

Recent studies proposed more complex methods that utilize the information extracted by wavelet analysis (Kan and Wong, 1998; Oussar and Dreyfus, 2000; Oussar et al., 1998; Wong and Leung, 1998; Xu and Ho, 2002; Zhang, 1997). These methods are not optimal, simply a trade-off between optimality and efficiency (He et al., 2002). Implementation of these methods can be summed up in the following three steps:

1. Construct a library W of wavelets.
2. Remove the wavelets whose support does not contain any sample points of the training data.
3. Rank the remaining wavelets and select the best wavelet regressors.

In the first step, the wavelet library can be constructed by either an orthogonal wavelet or a wavelet frame (He et al., 2002; Postalcioglu and Becerikli, 2007). By determining an orthogonal wavelet basis, the wavelet network is constructed simultaneously. However, to generate an orthogonal wavelet basis, the wavelet function has to satisfy strong restrictions (Daubechies, 1992; Mallat, 1999). In addition, the fact that orthogonal wavelets cannot be expressed in closed form makes them inappropriate for applications of function approximation or process modeling (Oussar and Dreyfus, 2000).

On the other hand, constructing wavelet frames is very easy and can be done by translating and dilating the mother wavelet selected. Results from Gao and Tsoukalas (2001) indicate that a family of compactly supported nonorthogonal wavelets is more appropriate for function approximation. Due to the fact that a wavelet family can contain a large number of wavelets, it is more convenient to use a truncated wavelet family than an orthogonal wavelet basis (Zhang, 1993).

However, constructing a wavelet network using wavelet frames is not a straightforward process. The wavelet library may contain a large number of wavelets since only the input data were considered in construction of the wavelet frame. To construct a wavelet network, the "best" wavelets must be selected. To do so, first the number of wavelet candidates must be decreased and the remaining wavelets must be ranked. The reduction of the wavelet candidates must be done carefully since arbitrary truncations may lead to large errors (Xu and Ho, 2005). In the second step, Zhang (1993) proposes removing wavelets that have very few training patterns in their support. Alternatively, magnitude-based methods were used by Cannon and Slotine (1995) to eliminate wavelets with small coefficients. In the third step, the remaining wavelets are ranked and the wavelets with the highest rank are used for construction of the wavelet network.

In the next section, three alternative methods are presented to reduce and rank the wavelets in the wavelet library: residual-based selection, stepwise selection by orthogonalization, and backward elimination.

Residual-Based Selection Initialization Method   In the framework of RBS, the wavelet that best fits the output data is selected first. Then the wavelet that best fits the residual of the fitting of the preceding stage is selected repeatedly. RBS is considered to be a very simple method but not an effective one (Juditsky et al., 1994). However, if the wavelet candidates reach a very large number, computational efficiency is essential and the RBS method may be used (Juditsky et al., 1994). Kan and Wong (1998) and Wong and Leung (1998) used the RBS algorithm for the synthesis of wavelet networks. Xu and Ho (2002) used a modified version of the RBS algorithm. An orthogonalized residual-based selection (ORBS) algorithm is proposed for more precise initialization of the wavelet network. The ORBS method combines the RBS and orthogonalized least squares (OLS) method. In this way, high efficiency is obtained while relatively low computational burden is maintained.

Selection by Orthogonalization Method The RBS method does not explicitly consider the interaction or nonorthogonality of the wavelets in the wavelet basis (Zhang, 1997). The SSO method is an extension of the RBS first proposed by Chen et al. (1989, 1991). The following procedure is followed to initialize the wavelet network: First, the wavelet that best fits the output data is selected. Then the wavelet that best fits the residual of the fitting of the previous stage together with the wavelet selected previously is selected repeatedly. In other words, the SSO considers the interaction or nonorthogonality of the wavelets. The selection of the wavelets is performed using the modified Gram–Schmidt algorithm, which has better numerical properties and is computationally less expensive than the ordinary Gram–Schmidt algorithm (Zhang, 1997). More precisely, the modified Gram–Schmidt algorithm is applied with a particular choice of the order in which the vectors are orthogonalized. SSO is considered to have good efficiency while not being expensive computationally. An algorithm similar to SSO was proposed by Oussar and Dreyfus (2000).

GRAM–SCHMIDT ALGORITHM The ranking method adopted here is the Gram–Schmidt procedure. The algorithm has been presented in detail by Chen et al. (1989). In this section the basic aspects of the algorithm are presented briefly. The basic problem that we want to solve is the following: Suppose that $\{x_1, x_2, \ldots, x_k\} \subset \mathbb{R}^n$ is a linear independent set of vectors. How do we find an orthonormal set of vectors $\{q_1, q_2, \ldots, q_k\}$ with the property span$\{q_1, q_2, \ldots, q_k\}$ = span$\{x_1, x_2, \ldots, x_k\}$?

A set of vectors are called *orthonormal* if the following two conditions are met:

$$\langle q_i, q_j \rangle = q_i \cdot q_j = 0 \qquad \text{for each } i \neq j \tag{3.11}$$

$$\|q_i\| = 1 \qquad \text{for all } i \tag{3.12}$$

Equation (3.11) ensures that the vectors q_i are orthogonal and (3.12) that they are normalized. Consider a model, linear with respect to its parameters, with k inputs and a training set of n training patterns. The inputs are ranked as follows. First, the input vector that has the smallest angle with the vector output is selected. Then all other input vectors, and the output vector, are projected onto the subspace orthogonal to the input vector selected. In this subspace of dimension $k - 1$, the procedure is iterated, and it is terminated when all inputs are ranked. That is, at step j the jth column is made orthogonal to each of the $j - 1$ columns orthogonalized previously, and the operations are repeated for $j = 2, \ldots, k$.

Since the output of the wavelet network is linear with respect to the weights, the procedure described above can easily be used where each input is actually the output of a multidimensional wavelet. Therefore, at the end of the procedure, all the wavelets of the library are ranked. The part of the process output not yet modeled and the wavelets not yet selected are subsequently projected onto the subspace orthogonal to the regressor selected. The procedure is repeated until all wavelets are ranked.

It is well known that the Gram–Schmidt procedure is very sensitive to round-off errors (Chen et al., 1989). It has been proved that if the regression matrix is ill-conditioned, the orthogonality will be lost. On the other hand, the modified Gram–Schmidt procedure is numerically stable. In this procedure, at the *j*th stage the columns subscripted $j + 1, \ldots, k$ are made orthogonal to the *j*th column and the operations repeated for $j = 1, \ldots, k - 1$. Both methods perform basically the same operations, only in a different sequence. However, the two algorithms have distinct differences in computational behavior. The modified procedure is more accurate and more stable than the classical algorithm. In all simulations presented below we have used the modified Gram–Schmidt method.

The algorithm can be summarized in the following two steps. At step 1 we must compute the orthogonal vectors $\{z_1, \ldots, z_k\}$. Hence, we set

$$z_1 = x_1$$

$$z_2 = x_2 - \mathrm{proj}_{z_1}(x_2) = x_2 - \frac{\langle x_2, z_1 \rangle}{\langle z_1, z_1 \rangle} z_1$$

$$z_k = x_k - \mathrm{proj}_{z_1}(x_k) - \cdots - \mathrm{proj}_{z_{k-1}}(x_k) = x_k - \frac{\langle x_k, z_1 \rangle}{\langle z_1, z_1 \rangle} z_1 - \cdots - \frac{\langle x_k, z_{k-1} \rangle}{\langle z_{k-1}, z_{k-1} \rangle} z_{k-1}$$

At step 2 the vectors $\{z_1, \ldots, z_k\}$ are normalized:

$$q_1 = \frac{z_1}{\|z_1\|}, \quad q_2 = \frac{z_2}{\|z_2\|}, \quad \ldots, \quad q_k = \frac{z_k}{\|z_k\|}$$

Backward Elimination Method In contrast to previous methods, the BE starts the regression by selecting all available wavelets from the wavelet library. The wavelet that contributes the least in the fitting of the training data is repeatedly eliminated. The objective is to increase the residual at each stage as little as possible. The drawback of BE is that it is computationally expensive, but it is considered to have good efficiency.

Zhang (1997) presented the exact number of arithmetic operations for each algorithm. More precisely, for the RSO at each step *i* the computational cost is $2nL - 2in + 6n - 1$, where *n* is the length of the training samples and *L* is the number of wavelets in the wavelet basis. Similarly, the computational cost of the SSO algorithm at each step is $8nL - 6in + 9n + 5L$. Roughly speaking, the SSO is four times more computationally expensive than the RSO algorithm. Finally, the operations needed in the BE method is $2n(L - i) + 4(L^2 + i^2) - 8iL - (L - i)$. In addition, at the beginning of the BE algorithm an $L \times L$ matrix must be inverted. If the number of hidden units, and as a result the number of wavelets that must be selected is $HUs > L/2$, fewer steps are performed by the BE algorithm, whereas for $HUs < L/2$ the contrary is true (Zhang, 1997).

All methods described above are used only for the initialization of the dilation and translation parameters. The network is trained further to obtain the vector of the parameters $w = \hat{\mathbf{w}}_n$, which minimizes the cost function. It is clear that additional computational burden is added to initialize the wavelet network efficiently. However, the efficient initialization significantly reduces the training phase; hence, the total number of computations is significantly smaller than in a network with random initialization.

Training a Wavelet Network with Backpropagation

After the initialization phase, the network is trained further to find the weights that minimize the cost function. As wavelet networks have gained in popularity, more complex training algorithms have been presented in the literature. Cristea et al. (2000) used genetic algorithms to train a wavelet network; Li and Chen (2002) proposed a learning algorithm that utilized least trimmed squares. He et al. (2002) suggest a hierarchical evolutionary algorithm. Xu and Ho (2005) employed the Levenberg–Marquardt algorithm. Chen et al. (2006) combine adaptive diversity learning particle swarm optimization and gradient descent algorithms to train a wavelet network. However, most evolutionary algorithms that include particle swarm optimization are inefficient and cannot avoid completely certain degeneracy and local minimum (Zhang, 2009). Also, evolutionary algorithms suffer from fine-tuning inefficiency (Chen et al., 2006; Yao, 1999). On the other hand, the Levenberg–Marquardt is one of the fastest algorithms for training classical sigmoid neural networks. The main drawback of this algorithm is that it requires the storage and inversion of some matrices that can be quite large.

The algorithms above originate from classical sigmoid neural networks, as they do not take advantage of the properties of wavelets (Zhang, 2007, 2009). Since a wavelet is a function whose energy is well localized in time frequency, Zhang (2007, 2009) used sampling theory to train a wavelet network in both uniform and nonuniform data. Their results indicate that the algorithm they proposed has global convergence.

In our implementation, ordinary backpropagation (BP) was used. Backpropagation is probably the most popular algorithm used for training wavelet networks (Fang and Chow, 2006; Jiao et al., 2001; Oussar and Dreyfus, 2000; Oussar et al., 1998; Postalcioglu and Becerikli, 2007; Zhang, 1997, 2007; Zhang and Benveniste, 1992). Ordinary BP is less fast but also less prone to sensitivity to initial conditions than are higher-order alternatives (Zapranis and Refenes, 1999).

The basic idea of backpropagation is to find the percentage contribution of each weight to the error. The error e_p for pattern p is simply the difference between the target (y_p) and the network output $(\hat{y}_p)$. By squaring and multiplying by $\frac{1}{2}$, we take the pairwise error E_p, which is used in network training:

$$E_p = \tfrac{1}{2}(y_p - \hat{y}_p)^2 = \tfrac{1}{2}e_p^2 \qquad (3.13)$$

The weights of the network were trained to minimize the mean quadratic cost function (or loss function):

$$L_n = \frac{1}{n}\sum_{p=1}^{n} E_p = \frac{1}{2n}\sum_{p=1}^{n} e_p^2 = \frac{1}{2n}\sum_{p=1}^{n}(y_p - \hat{y}_p)^2 \qquad (3.14)$$

Other functions can be used instead of (3.14); however, the mean quadratic cost function is the most commonly used. The network is trained until a vector of weights $w = \hat{\mathbf{w}}_n$ that minimizes the proposed cost function is found. The previous solution corresponds to a training sample of size n. Computing the parameter vector $\hat{\mathbf{w}}_n$ is always done by iterative methods. At each iteration t the derivative of the loss function with respect to the network weights is calculated. Then the parameters are updated using the following (delta) learning rule:

$$w_{t+1} = w_t - \eta \frac{\partial L_n}{\partial w_t} + \kappa(w_t - w_{t-1}) \tag{3.15}$$

where η is the learning rate and it is constant. The complete vector of the network parameters comprises $w = (w_i^{[0]}, w_{(\xi)ij}^{[1]}, w_{(\zeta)ij}^{[1]}, w_j^{[2]}, w_{\lambda+1}^{[2]})$.

A constant momentum term, defined by κ, is induced which increases the training speed and helps the algorithm to avoid oscillations. The learning rate and momentum speed take values between 0 and 1. Choice of the learning rate and the momentum depend on the application and the training sample. Usually, values between 0.1 and 0.4 are used.

The partial derivative of the cost function with respect to a weight w is given by

$$\frac{\partial L}{\partial w} = \frac{1}{2n} \sum_{p=1}^{n} \frac{\partial E_p}{\partial w} = \frac{1}{2n} \sum_{p=1}^{n} \frac{\partial E_p}{\partial \hat{y}_p} \frac{\partial \hat{y}_p}{\partial w}$$

$$= \frac{1}{n} \sum_{p=1}^{n} -(y_p - \hat{y}_p) \frac{\partial \hat{y}_p}{\partial w} = \frac{1}{n} \sum_{p=1}^{n} -e_p \frac{\partial \hat{y}_p}{\partial w} \tag{3.16}$$

The partial derivatives with respect to each parameter $\partial \hat{y}_p / \partial w$, and with respect to each input variable $\partial \hat{y}_p / \partial x_i$, are presented below. The partial derivative with respect to (w.r.t.) the bias term $w_{\lambda+1}^{[2]}$ is given by

$$\frac{\partial \hat{y}_p}{\partial w_{\lambda+1}^{[2]}} = 1 \tag{3.17}$$

Similarly, the partial derivatives w.r.t. the direct connections from the input variables $w_i^{[0]}$ to the output of the wavelet network $\hat{y}$ are given by

$$\frac{\partial \hat{y}_p}{\partial w_i^{[0]}} = x_i \qquad i = 1, \dots, m \tag{3.18}$$

The partial derivatives w.r.t. the linear connections from the wavelons $w_j^{[2]}$ to the output of the wavelet network are given by

$$\frac{\partial \hat{y}_p}{\partial w_j^{[2]}} = \Psi_j(\mathbf{x}) \qquad j = 1, \ldots, \lambda \tag{3.19}$$

The partial derivatives w.r.t. the translation $w_{(\xi)ij}^{[1]}$ are given by

$$\frac{\partial \hat{y}_p}{\partial w_{(\xi)ij}^{[1]}} = \frac{\partial \hat{y}_p}{\partial \Psi_j(\mathbf{x})} \frac{\partial \Psi_j(\mathbf{x})}{\partial \psi(z_{ij})} \frac{\partial \psi(z_{ij})}{\partial z_{ij}} \frac{\partial z_{ij}}{\partial w_{(\xi)ij}^{[1]}}$$

$$= w_j^{[2]} \psi(z_{1j}) \cdots \psi'(z_{ij}) \cdots \psi(z_{mj}) \frac{-1}{w_{(\zeta)ij}^{[1]}}$$

$$= -\frac{w_j^{[2]}}{w_{(\zeta)ij}^{[1]}} \psi(z_{1j}) \cdots \psi'(z_{ij}) \cdots \psi(z_{mj}) \tag{3.20}$$

The partial derivatives w.r.t. the dilation parameters $w_{(\zeta)ij}^{[1]}$ are given by

$$\frac{\partial \hat{y}_p}{\partial w_{(\zeta)ij}^{[1]}} = \frac{\hat{y}_p}{\partial \Psi_j(\mathbf{x})} \frac{\partial \Psi_j(\mathbf{x})}{\partial \psi(z_{ij})} \frac{\partial \psi(z_{ij})}{\partial z_{ij}} \frac{\partial z_{ij}}{\partial w_{(\zeta)ij}^{[1]}}$$

$$= w_j^{[2]} \psi(z_{1j}) \cdots \psi'(z_{ij}) \cdots \psi(z_{mj}) \frac{x_i - w_{(\xi)ij}^{[1]}}{w_{(\zeta)ij}^{[1]^2}}$$

$$= -\frac{w_j^{[2]}}{w_{(\zeta)ij}^{[1]}} \frac{x_i - w_{(\xi)ij}^{[1]}}{w_{(\zeta)ij}^{[1]^2}} \psi(z_{1j}) \cdots \psi'(z_{ij}) \cdots \psi(z_{mj})$$

$$= -\frac{w_j^{[2]}}{w_{(\zeta)ij}^{[1]}} z_{ij} \psi(z_{1j}) \cdots \psi'(z_{ij}) \cdots \psi(z_{mj})$$

$$= z_{ij} \frac{\partial \hat{y}_p}{\partial w_{(\xi)ij}^{[1]}} \tag{3.21}$$

Finally, the partial derivatives w.r.t. the input variables $\partial \hat{y}_p / \partial x_i$ are presented:

$$
\begin{aligned}
\frac{\partial \hat{y}_p}{\partial x_i} &= w_i^{[0]} + \frac{\displaystyle\sum_{j=1}^{\lambda} w_j^{[2]} \partial \Psi_j(\mathbf{x})}{\partial \psi(z_{ij})} \frac{\partial \psi(z_{ij})}{\partial z_{ij}} \frac{\partial z_{ij}}{\partial x_i} \\
&= w_i^{[0]} + \sum_{j=1}^{\lambda} w_j^{[2]} \psi(z_{1j}) \cdots \psi'(z_{ij}) \cdots \psi(z_{mj}) \frac{1}{w_{(\zeta)ij}^{[1]}} \\
&= w_i^{[0]} + \sum_{j=1}^{\lambda} \frac{w_j^{[2]}}{w_{(\zeta)ij}^{[1]}} \psi(z_{1j}) \cdots \psi'(z_{ij}) \cdots \psi(z_{mj}) \\
&= w_i^{[0]} - \sum_{j=1}^{\lambda} \frac{\partial \hat{y}_p}{\partial w_{(\xi)ij}^{[1]}}
\end{aligned}
\tag{3.22}
$$

The partial derivatives $\partial \hat{y}_p / \partial x_i$ are needed for the variable selection algorithm that is presented in the following chapters.

Online Training   The training methodology described above falls into the category of off-line training. This means that the weights of the networks are updated after all training patterns are presented to the network. Alternatively, one can use online training methods. In online methods the weights are changed after each presentation of a training pattern. For some problems, this method may yield effective results, especially for problems where data arrive in real time (Samarasinghe, 2006). Using online training it is possible to reduce training times significantly. However, for complex problems it is possible that online training will create a series of drawbacks. First, there is the possibility that the training will stop before the presentation of each training pattern to the network. Second, by changing the weights after each pattern, they could bounce back and forth with each iteration, possibly resulting in a substantial amount of wasted time (Samarasinghe, 2006). Hence, to ensure the stability of the algorithms, off-line training is used in this book.

Stopping Conditions for Training

After the initialization phase of the network parameters w, the weights $w_i^{[0]}$ and $w_j^{[2]}$ and the parameters $w_{(\xi)ij}^{[1]}$ and $w_{(\zeta)ij}^{[1]}$ are trained during the learning phase for approximating the target function. A key decision related to the training of a wavelet network involves when the weight adjustment should end. If the training phase stops early, the wavelet network will not be able to learn the underlying function of the training data and as a result will not perform well in predicting new, unseen data. On the other hand, if the training phase continues beyond the appropriate iterations, the network will begin to learn the noise part of the data and will become overfitted. As a result, the generalization ability of the network will be lost. Hence, it will not be appropriate to use the wavelet network in predicting future data.

In the next section a procedure for selecting the correct topology of a wavelet network is presented. Under the assumption that the wavelet network contains the number of wavelets that minimizes the prediction risk, the training is stopped when one of the following criteria is met: The cost function reaches a fixed lower bound or the variations of the gradient, the variations of the parameters reach a lower bound, or the number of iterations reaches a fixed maximum, whichever is satisfied first. In our implementation the fixed lower bound of the cost function, of the variations of the gradient, and of the variations of the parameters were set to 10^{-5}.

Evaluating the Initialization Methods

As mentioned earlier, the initialization phase is very important to the construction and training of a wavelet network. In this section we compare four different initialization methods: the heuristic, SSO, RBS, and BE methods, which constitute the bases for alternative algorithms and can be used with the BP training algorithm. The four initialization methods will be compared in three stages. First, the distance between the initialization and the underlying function as well as the training data will be measured. Second, the number of iterations needed to train the wavelet network will be compared. Finally, the difference between the final approximation of the trained wavelet network and the underlying function and the training data will be examined. The four initialization methods will be tested in two cases: on a simple underlying function and on a more complex function that incorporates large outliers.

Case 1: Sinusoid and Noise with Decreasing Variance   In the first case the underlying function $f(x)$ is given by

$$f(x) = 0.5 + 0.4 \sin 2\pi x + \varepsilon_1(x) \qquad x \in [0, 1] \tag{3.23}$$

where x is equally spaced in [0,1] and the noise $\varepsilon_1(x)$ follows a normal distribution with mean zero and decreasing variance:

$$\sigma_\varepsilon^2(x) = 0.05^2 + 0.1(1 - x^2) \tag{3.24}$$

The four initialization methods are examined using a wavelet network with 2 hidden units with learning rate 0.1 and momentum 0. The choice of the network structure proposed will be justified in Chapter 4. The training sample consists of 1.000 patterns.

Figure 3.2 shows the initialization of the four algorithms for the first training sample. It is clear that the heuristic algorithm produces the worst initialization. However, even the heuristic approximation is still better than a random initialization. On the other hand, initialization of the RBS algorithm gives a better approximation of the data; however, the approximation of the target function $f(x)$ is still not very

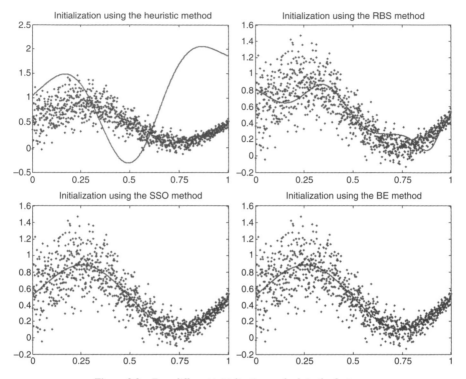

Figure 3.2   *Four different initialization methods in the first case.*

good. Finally, both the SSO and BE algorithms start very close to the target function $f(x)$. For the construction of the wavelet basis, the input dimension was scanned in four scale levels and 15 wavelet candidates were identified. To reduce the number of wavelet candidates, wavelets that contain fewer than three patterns in their support are removed from the basis. However, all wavelets contain at least three training samples; hence, the wavelet basis was not truncated.

The mean squared error (MSE) between the initialization of the network and the training data confirms the results cited above. More precisely, the MSE between the initialization of the network and the training data is 0.630809, 0.040453, 0.031331, and 0.031331 for the heuristic, RBS, SSO, and BE, respectively. Next we test how close the initialization is to the underlying function. The MSE between the initialization of the network and the underlying function is 0.59868, 0.302782, 0.000121, and 0.000121 for the heuristic, RBS, SSO, and BE, respectively. The results above indicate that the SSO and the BE produce the best initialization for the parameters of the wavelet network.

Another way to compare the initialization methods is to compare the number of iterations needed in the training phase until the solution $\hat{\mathbf{w}}_n$ is found. Also, whether or not the initialization methods proposed allows the training procedure to find the global minimum of the loss function will be examined.

The heuristic method was used to train 100 networks with different initial conditions for the direct connections $w_i^{[0]}$ and weights $w_j^{[2]}$. Training 100 networks with perturbed initial conditions is expected to be sufficient to avoid any possible local minimums of the loss function (3.14). It was found that the smallest MSE between the target function $f(x)$ and the final approximation of the network was 0.031331.

Using the RBS, the training phase stopped after 616 iterations. The overall fit was very good and the MSE between the network output and the training data was 0.031401, indicating that the network was stopped before the minimum of the loss function was achieved. Finally, the MSE between the network output and the target function was 0.000676.

On the other hand, when initializing the wavelet network with the SSO algorithm, only one iteration was needed in the training phase and the MSE was 0.031331, whereas the MSE between the underlying function $f(x)$ and the network approximation was only 0.000121. The same results were achieved using the BE method. Finally, one implementation of the heuristic method needed 1501 iterations. The results are presented in Table 3.1.

The results above indicate that the SSO and BE algorithms give the same results and significantly outperform both the heuristic and RBS algorithms. Moreover, the results above indicate that having a very good initialization not only significantly reduces the needed training iterations and as a result the total needed training time, but a vector of weights $\hat{\mathbf{w}}_n$ that minimizes the loss function can also be found.

Case 2: Sum of Sinusoids and Cauchy Noise Next, a more complex case is introduced where the function $g(x)$ is given by

$$g(x) = 0.5x \sin x + \cos^2 x + \varepsilon_2(x) \qquad x \in [-6, 6] \qquad (3.25)$$

TABLE 3.1 Initialization of the Four Methods[a]

	Heuristic	RBS	SSO	BE
Case 1				
MSE	0.031522	0.031401	0.031331	0.031331
MSE+	0.000791	0.000626	0.000121	0.000121
IMSE	0.630807	0.040453	0.031331	0.031331
IMSE+	0.598680	0.302782	0.000121	0.000121
Iterations	1501	616	1	1
Case 2				
MSE	0.106238	0.004730	0.004752	0.004364
MSE+	0.102569	0.000558	0.000490	0.000074
IMSE	7.877472	0.041256	0.012813	0.008403
IMSE+	7.872084	0.037844	0.008394	0.004015
Iterations	4433	3097	751	1107

[a]Case 1 refers to function $f(x)$ and case 2 to function $g(x)$. RBS, residual-based selection; SSO, stepwise selection by orthogonalization; BE, backward elimination; MSE, MSE between the training data and the network approximation; MSE+, MSE between the underlying function and the network approximation; IMSE, MSE between the training data and the network initialization; IMSE+, MSE between the underlying function and the network initialization.

and $\varepsilon_2(x)$ follows a Cauchy distribution with location 0 and scale 0.05, and x is equally spaced in $[-6, 6]$. The training sample consists of 1.000 training patterns. Whereas the first function is very simple, the second one, proposed by Li and Chen (2002), incorporates large outliers in the output space. The sensitivity to the presence of outliers of the wavelet network proposed will be tested. To approximate function $g(x)$, a wavelet network with 8 hidden units with learning rate 0.1 and momentum 0 is used. The choice of the topology proposed for the wavelet network is justified in Chapter 4.

The results obtained in the second case are similar. A closer inspection of Figure 3.3 reveals that the heuristic algorithm produces the worst initialization in approximating the underlying function $g(x)$. The RBS algorithm produces a significantly better initialization than the heuristic method; however, the initial approximation still differs from the training target values. Finally, both the BE and SSO algorithms produce a very good initialization. It is clear that the first approximation of the wavelet network is very close to the underlying function $g(x)$.

For construction of the wavelet basis, the input dimension was scanned in depth at five scale levels. The analysis revealed 31 wavelet candidates, and all of them were used for construction of the wavelet basis since all of them had at least three training samples in their support. The MSE between the initialization of the network and the

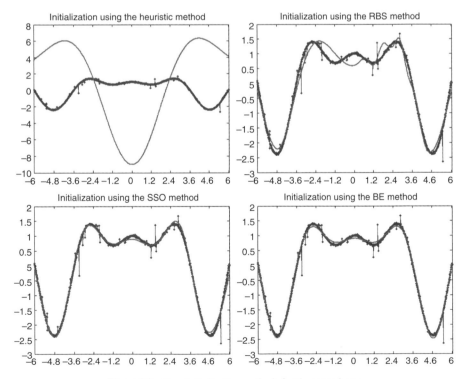

Figure 3.3 *Four initialization methods for the second case.*

training data was 7.87472, 0.041256, 0.012813, and 0.008403 for the heuristic, RBS, SSO, and BE algorithms, respectively. Also, the MSE between the initialization of the network and the underlying function $g(x)$ was 7.872084, 0.037844, 0.008394, and 0.004015 for the heuristic, RBS, SSO, and BE, respectively. The previous results indicate that the training phase using the BE algorithm starts very close to the target function $g(x)$.

Next, the number of iterations needed in the training phase of each method was compared. Also, whether or not the initialization methods proposed allow the training procedure to find the global minimum of the loss function was examined. The RBS algorithm stopped after 3097 iterations, and the MSE of the final approximation of the wavelet network and the training patterns was 0.004730. The MSE between the underlying function $f(x)$ and the network approximation was 0.000558. When initializing the wavelet network with the SSO algorithm only, 741 iterations were needed in the training phase and the MSE was 0.004752, while the MSE between the underlying function $g(x)$ and the network approximation was 0.000490. The BE needed 1107 iterations in the training phase and the MSE was 0.004364, while the MSE between the underlying function $g(x)$ and the network approximation was only 0.000074. Finally, one implementation of the heuristic method needed 4433 iterations and the MSE was 0.106238, while the MSE between the underlying function $g(x)$ and the network approximation was 0.102569. The results are presented in the second part of Table 3.1. In the second case the BE was slower than the SSO; however, the final approximation was significantly closer to the target function than with any other method. Note that in all cases the training was stopped when the minimum velocity, 10^{-5}, was reached. If the minimum velocity is reduced further, the SSO algorithm will produce results similar to those of the BE, but extra computational time will be needed.

The previous examples indicate that the SSO and BE perform similarly and outperform the other two methods, whereas BE outperforms the SSO in complex problems. Previous studies suggest that the BE is more efficient than the SSO algorithm; however, it is more computationally expensive. On the other hand, in the BE algorithm it is necessary to calculate the inverse of the wavelet matrix, whose columns might be linearly dependent (Zhang, 1997). In that case the SSO must be used. However, since the wavelets come from a wavelet frame, this happens very rarely (Zhang, 1997).

CONCLUSIONS

In this chapter the structure of the wavelet network proposed was presented and discussed analytically. A wavelet network usually has the form of a three-layer network. The lower layer represents the input layer, the middle layer is the hidden layer, and the upper layer is the output layer. In our implementation, a multidimensional wavelet network with a linear connection between the wavelons and the output is utilized. In addition, there are direct connections from the input layer to the output layer that will help the network to perform well in linear applications.

Furthermore, the initialization phase, training phase, and stopping conditions are discussed. The initialization of the parameters is very important in wavelet networks since it can reduce the training time significantly. The initialization method developed extracts useful information from the wavelet analysis. The simplest method is the heuristic method; more sophisticated methods, such as residual-based selection, selection by orthogonalization, and backward elimination can be used for efficient initialization.

The results from our analysis indicate that backward elimination significantly outperforms other methods. The results from two simulated cases show that using the backward elimination method, the wavelet network provides a fit very close to the real underlying function. However, it is more computationally expensive than the other methods.

The backpropagation method is used for network training. Iteratively, the weights of the networks are updated based on the delta learning rule, where a learning rate and a momentum are used. The weights of the network were trained to minimize the mean quadratic cost function. The training continues until one of the stopping conditions is met.

REFERENCES

Anders, U., and Korn, O. (1999). "Model selection in neural networks." *Neural Networks*, 12(2), 309–323.

Becerikli, Y. (2004). "On three intelligent systems: dynamic neural, fuzzy and wavelet networks for training trajectory." *Neural Computation and Applications*, 13, 339–351.

Becerikli, Y., Oysal, Y., and Konar, A. F. (2003). "On a dynamic wavelet network and its modeling application." *Lecture Notes in Computer Science*, 2714, 710–718.

Billings, S. A., and Wei, H.-L. (2005). "A new class of wavelet networks for nonlinear system identification." *IEEE Transactions on Neural Networks*, 16(4), 862–874.

Cannon, M., and Slotine, J.-J. E. (1995). "Space–frequency localized basis function networks for nonlinear system estimation and control." *Neurocomputing*, 9, 293–342.

Chen, Y., Billings, S. A., and Luo, W. (1989). "Orthogonal least squares methods and their application to non-linear system identifcation." *International Journal of Control*, 50, 1873–1896.

Chen, Y., Cowan, C., and Grant, P. (1991). "Orthogonal least squares learning algorithm for radial basis function networks." *IEEE Transactions On Neural Networks*, 2, 302–309.

Chen, Y., Yang, B., and Dong, J. (2006). "Time-series prediction using a local linear wavelet neural wavelet." *Neurocomputing*, 69, 449–465.

Cristea, P., Tuduce, R., and Cristea, A. (2000). "Time series prediction with wavelet neural networks." *Proceedings of 5th Seminar on Neural Network Applications in Electrical Engineering*, Belgrade, Yugoslavia, 5–10.

Daubechies, I. (1992). *Ten Lectures on Wavelets*. SIAM, Philadelphia.

Fang, Y., and Chow, T. W. S. (2006). "Wavelets based neural network for function approximation." *Lecture Notes in Computer Science*, 3971, 80–85.

Gao, R., and Tsoukalas, H. I. (2001). "Neural-wavelet methodology for load forecasting." *Journal of Intelligent and Robotic Systems*, 31, 149–157.

He, Y., Chu, F., and Zhong, B. (2002). "A hierarchical evolutionary algorithm for constructing and training wavelet networks." *Neural Computing and Applications*, 10, 357–366.

Jiao, L., Pan, J., and Fang, Y. (2001). "Multiwavelet neural network and its approximation properties." *IEEE Transactions on Neural Networks*, 12(5), 1060–1066.

Juditsky, A., Zhang, Q., Delyon, B., Glorennec, P.-Y., and Benveniste, A. (1994). "Wavelets in identification—wavelets, splines, neurons, fuzzies: how good for identification?" Techincal Report, INRIA.

Kadambe, S., and Srinivasan, P. (2006). "Adaptive wavelets for signal classification and compression." *International Journal of Electronics and Communications*, 60, 45–55.

Kan, K.-C., and Wong, K. W. (1998). "Self-construction algorithm for synthesis of wavelet networks." *Electronic Letters*, 34, 1953–1955.

Khayamian, T., Ensafi, A. A., Tabaraki, R., and Esteki, M. (2005). "Principal component-wavelet networks as a new multivariate calibration model." *Analytical Letters*, 38(9), 1447–1489.

Li, S. T., and Chen, S.-C. (2002). "Function approximation using robust wavelet neural networks." *Proceedings of ICTAI '02*, Washington, DC, 483–488.

Mallat, S. G. (1999). *A Wavelet Tour of Signal Processing*. Academic Press, San Diego, CA.

Mellit, A., Benghamen, M., and Kalogirou, S. A. (2006). "An adaptive wavelet-network model for forecasting daily total solar-radiation." *Applied Energy*, 83, 705–722.

Oussar, Y., and Dreyfus, G. (2000). "Initialization by selection for wavelet network training." *Neurocomputing*, 34, 131–143.

Oussar, Y., Rivals, I., Presonnaz, L., and Dreyfus, G. (1998). "Trainning wavelet networks for nonlinear dynamic input output modelling." *Neurocomputing*, 20, 173–188.

Postalcioglu, S., and Becerikli, Y. (2007). "Wavelet networks for nonlinear system modelling." *Neural Computing and Applications*, 16, 434–441.

Samarasinghe, S. (2006). *Neural Networks for Applied Sciences and Engineering*. Taylor & Francis Group, New York.

Wong, K.-W., and Leung, A. C.-S. (1998). "On-line successive synthesis of wavelet networks." *Neural Processing Letters*, 7, 91–100.

Xu, J., and Ho, D. W. C. (1999). "Adaptive wavelet networks for nonlinear system identification." *Proceedings of American Control Conference*, San Diego, CA.

Xu, J., and Ho, D. W. C. (2002). "A basis selection algorithm for wavelet neural networks." *Neurocomputing*, 48, 681–689.

Xu, J., and Ho, D. W. C. (2005). "A constructive algorithm for wavelet neural networks." *Lecture Notes in Computer Science*, 3610, 730–739.

Yao, X. (1999). "Evolving artificial neural networks." *Proceedings IEEE*, 87(9), 1423–1447.

Zapranis, A., and Refenes, A. P. (1999). *Principles of Neural Model Indentification, Selection and Adequacy: With Applications to Financial Econometrics*. Springer-Verlag, New York.

Zhang, Q. (1993). "Regressor selection and wavelet network construction." Technical Report, INRIA.

Zhang, Q. (1994). "Using Wavelet Network in Nonparametric Estimation." Technical Report, 2321, INRIA.

Zhang, Q. (1997). "Using wavelet network in nonparametric estimation." *IEEE Transactions on Neural Networks*, 8(2), 227–236.

Zhang, Q., and Benveniste, A. (1992). "Wavelet networks." *IEEE Transactions on Neural Networks*, 3(6), 889–898.

Zhang, Z. (2007). "Learning algorithm of wavelet network based on sampling theory." *Neurocomputing*, 71(1), 224–269.

Zhang, Z. (2009). "Iterative algorithm of wavelet network learning from nonuniform data." *Neurocomputing*, 72, 2979–2999.

Zhao, J., Chen, B., and Shen, J. (1998). "Multidimensional non-orthogonal wavelet-sigmoid basis function neural network for dynamic process fault diagnosis." *Computers and Chemical Engineering*, 23, 83–92.

4

Model Selection: Selecting the Architecture of the Network

In this chapter we describe the model selection procedure. One of the most crucial steps is to identify the correct topology of the network. A desired wavelet network architecture should contain as few hidden units as necessary; at the same time it should explain as much variability of the training data as possible. A network with fewer hidden units than needed would not be able to learn the underlying function; selecting more hidden units than needed would result in an overfitted model. Therefore, it is essential to derive an algorithm to select the appropriate wavelet network model for a given problem.

The simplest way to select the optimal number of hidden units—in other words, the architecture of the wavelet network—is by trial and error, a method called *exhaustive search*. To do so, the training patterns must be split into a training sample and a validation sample. This method suggests that the optimum number of wavelons is given by the structure of the wavelet network that gives the minimum error on the validation set. This method is very simple but also very time consuming, and the information in the validation sample is not utilized for better training of the wavelet network.

In *early stopping* a fixed and large number of hidden units is used in construction of the network. The weights are allowed to change during the training phase. These free parameters are growing during the training phase. In early stopping the training is stopped to avoid overfitting of the wavelet network to the data.

Wavelet Neural Networks: With Applications in Financial Engineering, Chaos, and Classification, First Edition. Antonios K. Alexandridis and Achilleas D. Zapranis.
© 2014 John Wiley & Sons, Inc. Published 2014 by John Wiley & Sons, Inc.

Another approach to avoid overfitting is *regularization*. In regularization, a parameter larger than zero is specified, and a regularized performance index is minimized instead of the original mean squared error. The idea is to keep the overall growth of weights to a minimum such that weights are pulled toward zero. In this process, only the important weights are allowed to grow; the others are forced to decay.

Both early stopping and regularization use all weights in training. As a result, the structural complexity of the network is not reduced. Alternatively, in the *pruning method* the complexity of the network is reduced so that only the essential weights and neurons remain in the model. However, the various criteria used in pruning the weights are not employed in a statistical way.

Finally, another method used to find the optimal architecture of a wavelet network is *minimum prediction risk* (MPR). The idea behind MPR is to estimate the out-of-sample performance of incrementally growing networks. The number of hidden units that minimizes the prediction risk is the appropriate number of hidden units that should be used for construction of a wavelet network. In other words, the prediction risk is a form of measurement of the generalization ability of the wavelet network.

Early methods used *information criteria* to estimate the minimum prediction risk. The most popular information criteria are Akaike's final prediction error (FPE), generalized cross-validation (GCV), and the Bayesian information criterion (BIC). Information criteria measure the error between the training data and the network output, but a penalty term is added for large networks. Information criteria were derived for linear models, and some of the assumptions that they employ are not necessarily true for nonlinear nonparametric wavelet networks.

Alternatively, *resampling schemes* such as *bootstrap* or *cross-validation* can be used. In resampling schemes, different versions of a single statistic that would ordinarily be calculated from one sample can be estimated. Bootstrap and cross-validation do not require prior identification of the data-generating process. The main disadvantage of the application of sampling techniques is the fact that they are computationally expensive.

THE USUAL PRACTICE

The usual approaches proposed in the literature are early stopping, regularization, and pruning. In early stopping a fixed and large number of hidden units is used in construction of the network. In regularization methods the weights of the network are trained to minimize the loss function plus a penalty term. The idea is to keep the overall growth of weights to a minimum such that weights are pulled toward zero. Therefore, only a subset of weights that become most sensitive to the output is used effectively. In this process, only the important weights are allowed to grow; others are forced to decay.

Early Stopping

In early stopping a fixed and large number of hidden units are used in construction of the network. Hence, a large number of weights must be initialized and optimized during the training phase. The number of weights roughly defines the degrees of freedom of the network. If the training phase continues past the appropriate number of iterations and the weights grow very large in the training phase, the network will begin to learn the noise part of the data and will become overfitted. As a result, the generalization ability of the network will be lost. Hence, it is not appropriate to use the wavelet network in predicting new and unseen data. On the other hand, if the training is stopped at an appropriate point, it is possible to avoid overfitting the network.

A common practice to overcome the problems outlined above is the use of a validation sample. At each iteration, the network is trained using the training sample. Then the cost function between the training data and the network output is estimated and it is used to adjust the weights. Then the generalization ability of the network is measured using the validation sample. More precisely, the network is used to forecast the target values of the validation sample using the unseen input data of the validation sample. The error between the network output and the target data of the validation sample is calculated. Usually, the validation sample has 10 to 30% the size of the training sample.

At the beginning of the training phase the errors of both the training and the validation sample will start to decrease as the network weights are adjusted to the training data. After a particular iteration the network will begin to learn the noise part of the data. As a result, the error of the validation sample will begin to increase. This is an indication that the network is starting to lose its generalization ability and that the training phase must be stopped.

In the early stopping method a more complex model than needed is used. Hence, a large number of weights must be trained. As a result, large training times are expected. Moreover, the network incorporates a large number of connections, most of them with small weights. In addition, a validation sample should be used. However, in real applications, usually only a small amount of data is available, and splitting the data is not useful. Furthermore, growing validation errors indicate the reduction of a network's complexity (Anders and Korn, 1999). Finally, the solution $\hat{w}_n$ of the network is highly dependent on dividing the data and the initial conditions (Dimopoulos et al., 1995).

Regularization

Another approach to avoiding overfitting is regularization. In regularization methods the weights of the network are trained to minimize the loss function plus a penalty term. Regularization is attempting to keep the overall growth of weights to a minimum by allowing only the important weights to grow. The remaining weights are pulled toward zero (Samarasinghe, 2006). This method is often called *weight decay* (Samarasinghe, 2006).

The regularization method tries to minimizes the sum:

$$W = L_n + \delta \sum_{j=1}^{J} w_j^2 \tag{4.1}$$

where the second term is the penalty term, w_j is a weight, J is the total number of weights in the network architecture, and δ is a regularization parameter. The penalty term is not restricted to the choice above. However, the penalty terms are usually chosen arbitrarily without theoretical justification (Anders and Korn, 1999). Moreover, a bad regularization parameter, δ, can severely restrict the growth of weights, and as result, the network will be underfitted (Samarasinghe, 2006).

Pruning

Similar to other methods, the aim of pruning is to identify those parameters that contribute the least to network performance. Several approaches have been proposed to prune networks. However, the significance of each weight is usually not measured in a statistical way (Anders and Korn, 1999). Reed (1993) has provided an extensive survey of pruning methods. One of the disadvantages of pruning is that it often does not take correlated weights into account. Two weights that cancel each other out do not have any effect at the output of the network; however, each weight may have a large effect (Reed, 1993). Also, the time when the pruning should stop is usually arbitrary (Reed, 1993). Reed (1993) separated the pruning algorithms into two major groups: sensitivity calculation methods and penalty term methods. Here we present selected methods from each group.

Brute-Force Pruning   In brute-force pruning, the simplest method, each weight is set to zero and the effect on the error is estimated. If the change in the error increases "too much," the weight is restored to its value. One way to do so is, first, to estimate the change in error for every weight and for every pattern and then delete the weight with the least effect. This procedure is repeated up to a fixed threshold. However, it is not very straightforward to define whether or not the increase in error is large.

Sensitivity Calculation: Saliency   Reed (1993) has estimated the saliency of a weight using the second derivative of the error with respect to the weight:

$$\delta L_n = \sum_{i=1}^{p} g_i \, \delta w_i + \frac{1}{2} \sum_{i=1}^{p} a_{ii} \, \delta w_i^2 + \frac{1}{2} \sum_{i=1,i\neq j}^{p} a_{ij} \, \delta w_i \, \delta w_j + O\left(\|\delta W\|^2\right) \quad (4.2)$$

where the δw_i's are the components of δW, g_i are the components of the gradient of L_n with respect to W, and the a_{ij} are the elements of the Hessian matrix H:

$$g_i = \frac{\partial L_n}{\partial w_i} \quad (4.3)$$

$$a_{ij} = \frac{\partial^2 L_n}{\partial w_i \, \partial w_j} \quad (4.4)$$

Since pruning is done on a well-defined local minimum and for small perturbations, (4.2) can be simplified since the Hessian matrix is very large:

$$\delta L_n \approx \frac{1}{2} \sum_{i=1}^{p} a_{ii} \delta w_i^2 \tag{4.5}$$

Then the saliency of weight w_k is given by

$$S(w_i) = \frac{a_{ii} w_i^2}{2} \tag{4.6}$$

It can be considered that a_{ii} is an indication of the acceleration of the error with respect to a small perturbation to a weight w_i. Hence, through equation (4.6), an indication of the total effect of w_i on the error is obtained. The larger the saliency, the larger the influence of w_i on error. The other entries of the Hessian matrix are assumed to be zero; therefore, the second derivative with respect to weights other than itself is ignored (Samarasinghe, 2006). This implies that the weights of the network are independent. However, this may not be true for a network that has more than the optimum number of weights.

To apply this method, the following procedure is followed. First, a wavelet network should be trained in the normal way and the saliency computed for each weight. Then, weights with small values of saliency are removed. This may lead to pruning of weights as well as neurons. After a removal of a weight, the wavelet network is trained further. The simplified trained wavelet network should perform as well as the optimum network with a larger number of weights (Samarasinghe, 2006).

Irrelevant Connection Elimination Scheme An extension of sensitivity calculation is the irrelevant connection elimination scheme proposed by Zapranis and Haramis (2001). Once the parameters of the wavelet neural model $g_\lambda (\mathbf{x}; \mathbf{w})$ are estimated, we have to deal with the presence of flat minima (potentially, many combinations of the network parameters corresponding to the same level of the empirical loss), especially if the statistical properties of the model are of importance, as is the case in complex financial applications. To identify a locally unique solution, we have to remove all the irrelevant parameters: that is, the parameters that do not affect the level of the empirical loss.

For this purpose we use the irrelevant connection elimination (ICE) scheme, which is much less computationally demanding than other alternatives. The irrelevant connection elimination scheme, although it uses the full Hessian of L_n, does not require inverting the Hessian matrix, a common requirement of other algorithms. ICE is based on Taylor's approximation of the empirical loss:

$$\delta L_n = \sum_{i=1}^{p} g_i \, \delta w_i + \frac{1}{2} \sum_{i=1}^{p} a_{ii} \, \delta w_i^2 + \frac{1}{2} \sum_{i=1, i \neq j}^{p} a_{ij} \, \delta w_i \, \delta w_j + O\left((\delta w)^3\right) \tag{4.7}$$

where

$$a_{ij} = \frac{\partial^2 L_n}{\partial w_i \, \partial w_j} \tag{4.8}$$

From (4.7), ICE derives the "saliencies" $S(w_i)$ (i.e., the contribution of w_i to δL_n) when a small perturbation δw_k is added to all connections:

$$S(w_i) = g_i \, \delta w_i + \frac{1}{2} \sum_{j=1}^{p} a_{ij} \, \delta w_i \, \delta w_j \tag{4.9}$$

where $\delta L_n = \sum_{i=1}^{p} S(w_i)$.

At a well-defined local minimum, (4.9) can be simplified by setting $g_i = 0$, although this is not a requirement. The method can be summarized in the following steps:

Step 1: Train to convergence.

Step 2: Compute the saliencies $S(w_i)$.

Step 3: Deactivate the connection with the least associated saliency, unless it was reactivated in step 5. When a prespecified maximum number of steps has been reached, the algorithm STOPS.

Step 4: Train further for a small number of epochs, until the training error has stabilized.

Step 5: If the training error has increased, reactivate the connection; otherwise, remove it. Then go to step 3.

Because of possible dependencies in the connections, it is not advisable to remove more than one connection at a time (the removal of one connection can affect the standard errors and saliencies of others). This does not pose any computational problems to ICE, since computing the Hessian is of the same order of complexity as computing the derivatives $\partial L_n / \partial w_i$ during training.

MINIMUM PREDICTION RISK

The aim of model selection is to find the least complex model that can learn the underlying target function. Previous methods do not use the optimal architecture of a wavelet network. A very large wavelet network is used and then various methods are developed to avoid overfitting. Smaller networks usually are faster to train and need less computational power to build (Reed, 1993).

Alternatively, the minimum prediction risk principle can be applied (Efron and Tibshirani, 1993; Zapranis and Refenes, 1999). The idea behind minimum prediction risk is to estimate the out-of-sample performance of incrementally growing networks. Assuming that the explanatory variables **x** were selected correctly and remain fixed, the model selection procedure is the following: The procedure starts with a fully connected network with no hidden units (in our proposed structure of wavelet networks, this is a linear model). The wavelet network is trained, and then the prediction risk is estimated. Then, iteratively, a new hidden unit is added to the network. The new wavelet networks are trained and the new prediction risk is estimated at each step. The number of hidden units that minimizes the prediction risk is the appropriate number of hidden units that should be used for construction of the wavelet network.

The prediction risk measures the generalization ability of the network. More precisely, the prediction risk of a network $g_\lambda(\mathbf{x}; \hat{\mathbf{w}}_n)$ is the expected performance of the network on new data that were not introduced during the training phase and is given by

$$P_\lambda = E\left[\frac{1}{n}\sum_{p=1}^{n}\left(y_p^* - \hat{y}_p^*\right)^2\right] \tag{4.10}$$

where $(\mathbf{x}_p^*, y_p^*)$ are the new observations that have not been used in the construction of the network $g_\lambda(\mathbf{x}; \hat{\mathbf{w}}_n)$, and $\hat{y}_p^*$ is the network output using the new observations, $g_\lambda(\mathbf{x}^*; \mathbf{w})$.

Finding a statistical measure that estimates the prediction risk is not a straightforward procedure, however. Since there is a linear relationship between the wavelons and the output of the wavelet network, Zhang (1993, 1994, 1997) and Zhang and Benveniste (1992) propose the use of information criteria widely applied previously in linear models. A different approach was presented by Zapranis and Refenes (1999). An analytical form of the prediction risk (4.10) was presented for sigmoid neural networks. However, the assumptions made by Zapranis and Refenes (1999) are not necessarily true in the framework of wavelet networks, and analytical forms are not available for estimating the prediction risk for wavelet networks. Alternatively, the use of sampling methods such as bootstrap and cross-validation can be employed since they do not depend on any assumptions regarding the model (Efron and Tibshirani, 1993). The only assumption made by sampling methods is that the data are a sequence of independent, identically distributed variables.

ESTIMATING THE PREDICTION RISK USING INFORMATION CRITERIA

In wavelet networks, the wavelons are connected linearly to the output of the wavelet network, Zhang (1993, 1994, 1997) and Zhang and Benveniste (1992) propose the use of information criteria to find the optimal architecture of a wavelet network. Information criteria are used widely and successfully in estimation of the number of parameters in linear models. More precisely, Zhang (1994) suggested that Akaike's final prediction error (FPE) can be used in various applications. More recently, Zhang (1997) suggested that generalized cross-validation (GCV) is an accurate tool for selecting the number of wavelets that constitutes the wavelet network's topology.

To estimate the prediction risk and to find the network with the best predicting ability, a series of information criteria was developed. As the model complexity increases and more parameters are added to the wavelet network, it is expected that the fit will improve, but not necessarily the forecasting ability of the wavelet network. The idea behind these criteria is to measure the error between the training data and the network output, but at the same time to penalize the complexity of the network.

To select the best architecture of the wavelet network, the following procedure is pursued. In the first step, a wavelet network with no hidden units is constructed. The wavelet network is trained and then the corresponding information criterion and the prediction risk are estimated. In the next step, 1 hidden unit is added to the architecture of the wavelet network and the procedure is repeated until the network contains a predefined maximum number of hidden units. The number of hidden units that produces the smallest prediction risk is the number of the appropriate wavelets for the construction of the wavelet network.

Akaike's Information Criterion   Several criteria exist for model selection. Early studies make use of the generalized prediction error (GPE) proposed by Moody (1992) and the network information criterion (NIC) proposed by Murata et al. (1994). However, the results by Anders and Korn (1999) indicate that NIC significantly underperforms other criteria. Alternatively, Akaike's information criterion (AIC) (Akaike, 1973, 1974), which was proved to work well in various cases, was used. AIC is given by

$$J_{\text{AIC}} = 2k + n \ln \left[\frac{1}{n} \sum_{p=1}^{n} \left(y_p - \hat{y}_p \right)^2 \right] \tag{4.11}$$

where k is the number of parameters of the network and n is the number of training patterns in the training sample. The target value is given by y_p, and $\hat{y}_p$ is the approximation of the target value by the network.

Final Prediction Error   Zhang (1994) suggested that Akaike's final prediction error (FPE) can be used in various applications. The FPE is given by

$$J_{\text{FPE}} = \frac{1 + k/n}{2n - 2k} \sum_{p=1}^{n} \left(y_p - \hat{y}_p \right)^2 \tag{4.12}$$

Generalized Cross-Validation   More recently, Zhang (1997) suggested that generalized cross-validation (GCV) should be used to select the number of wavelets that constitutes the wavelet network topology. GCV is given by

$$J_{\text{GCV}} = \frac{1}{n} \sum_{p=1}^{n} \left(y_p - \hat{y}_p \right)^2 + \frac{2\text{HU} \cdot \hat{\sigma}^2}{n} \tag{4.13}$$

In practice, the noise variance σ^2 is not known. In that case it has to be estimated. An estimate is given by the MSE between the network output and the target data (Zhang, 1997).

Bayesian Information Criterion Similar to GCV is the Bayesian information criterion (BIC), given by

$$J_{\text{BIC}} = \frac{1}{n} \sum_{p=1}^{n} \left(y_p - \hat{y}_p \right)^2 + \frac{k\hat{\sigma}^2 \ln(n)}{n} \tag{4.14}$$

To estimate the AIC, FPE, GCV, and BIC, the number of the hidden units is needed. Because we do not have a priori knowledge of the correct number of hidden units or parameters of a wavelet network for estimation of GCV and the BIC, we estimate the criteria above iteratively. The computational cost of these algorithms can be expressed as a function of wavelet networks to be trained. For example, to estimate the prediction risk using the FPE or GCV from 1 to five hidden units, five wavelet networks must be initialized and fully trained. The model selection algorithm using IC is illustrated in Figure 4.1. It is expected that the prediction risk will decrease (almost) monotonically until it reaches a minimum and then it will increase (almost) monotonically. The number of wavelons needed for the construction of the networks is the number of hidden units that minimize the prediction risk.

The criteria described above for estimation of the prediction risk are derived from linear models. Usually, these methods are based on assumptions that are not necessarily true in the framework of nonlinear nonparametric estimation. The hypothesis behind these information criteria is the asymptotic normality of the maximum likelihood estimators; hence, the information criteria are not theoretically justified for overparameterized networks (Anders and Korn, 1999).

Moreover, in fitting problems more complex than least squares, the number of parameters k is not known (Efron and Tibshirani, 1993) and it is unclear how to compute the degrees of freedom (Curry and Morgan, 2006) or the effective number of parameters described by Moody (1992).

Alternatively, the use of sampling methods such as bootstrapping and cross-validation was suggested (Efron and Tibshirani, 1993). The only assumption made by sampling methods is that the data are a sequence of independent, identically distributed variables. Bootstrapping and cross-validation do not require knowledge of the number of parameters k. Another advantage of bootstrapping and cross-validation is their robustness. In contrast to sampling methods, both GCV and BIC require a roughly correct model to obtain an estimate of the noise variance.

ESTIMATING THE PREDICTION RISK USING SAMPLING TECHNIQUES

Instead of using information criteria to obtain an estimate of the prediction risk, resampling schemes can be used. Two resampling schemes described in this section

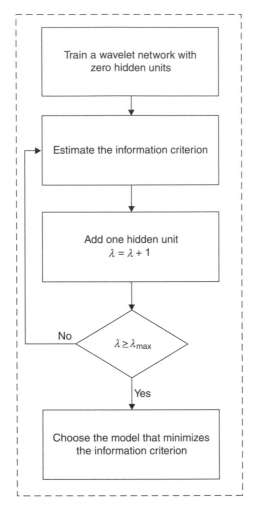

Figure 4.1   *Model selection algorithm using information criteria.*

are bootstrapping and cross-validation, which do not require prior identification of the data-generating process. Furthermore, they can approximate the distributional properties of a small sample $\hat{\mathbf{w}}_n$ accurately when the data are a sequence of independent, identically distributed variables. The estimation of these two approaches is asymptotically equal. The main disadvantage of the application of sampling techniques is the fact that they are computationally expensive.

Bootstrapping allows one to gather many alternative versions of a single statistic that would ordinarily be calculated from one sample. Where a set of observations can be assumed to originate from an independent, identically distributed population, bootstrapping can be implemented by constructing a number of new samples of the data set observed (and equal in size to the data set), each of which is obtained by random sampling with replacement from the original data set.

Similarly, in cross-validation, new samples are constructed from the original data set. However, there are two differences. First, random sampling is performed without replacement, and second, each sample is split into two parts, the training set and the validation set.

Bootstrapping

In this section the bootstrapping method is described. In summary, the simple bootstrapping approach generates new random instances from the original data. Then a new model is estimated for each sample, and finally, each fitted model is applied in order to generate an estimate of the prediction risk.

There are two methods of applying bootstrapping: bootstrapping pairs and bootstrapping residuals. In this book the bootstrapping pairs method is followed. The bootstrapping pairs method is less sensitive to assumptions than are bootstrapping residuals (Efron and Tibshirani, 1993). The only assumption behind bootstrapping pairs is that the original pairs were sampled randomly from some distribution. On the other hand, in bootstrapping residuals, the distribution of the residuals must be assumed beforehand. This is a very strong assumption which may lead to false conclusions.

Typically, a large number B of new samples $D_n^{*(b)} = \{x_p^{*(b)}, y_p^{*(b)}\}_{p=1}^n$ is created from the original sample, $D_n = \{x_p, y_p\}_{p=1}^n$, with size n, where $b = 1, \dots, B$. Typically, the number of samples is $20 < B$. Each pattern $\{x_p, y_p\}$ has $1/n$ probability to be selected with replacement from the original sample.

Next, we illustrate how bootstrapping works by presenting a very simple example. Let us assume that the variable $x = \{1, 2, \dots, 10\}$ and that the dependent variable y is given by the relationship $y_p = x_p^2$. Hence, $y = \{1, 4, \dots, 100\}$. The original data set consists of 10 patterns $\{x_p, y_p\}$. By applying bootstrapping we create 10 new samples of size 10 each. An example of 10 bootstrapped samples is presented in Table 4.1.

For each new sample $D_n^{*(b)}$, the wavelet network is trained to find the weight vector $\hat{\mathbf{w}}_n^{*(b)}$, and the loss function $L_n(\hat{\mathbf{w}}_n^{*(b)})$ is estimated. Then an estimation of the prediction risk is given by

$$\hat{P}_\lambda = \frac{1}{nB} \sum_{b=1}^B \sum_{p=1}^n \left\{ y_p - g_\lambda \left(\mathbf{x}_p; \hat{\mathbf{w}}_n^{*(b)} \right) \right\}^2 \tag{4.15}$$

The previous formula for the prediction risk is the average performance of the wavelet networks, which were trained on bootstrapped samples, on the original data set. In other words, it is the average of all loss functions estimated of the bootstrapped wavelet networks to the original sample.

In the estimation of prediction risk above, the wavelet network from each bootstrapped sample was used to predict the target values of the original sample. The estimation above of the prediction risk is very simple in use; however, it is known that it is not very accurate (Efron and Tibshirani, 1993).

TABLE 4.1   Simple Bootstrapping Example

	Pattern									
Sample	1	2	3	4	5	6	7	8	9	10
$\mathbf{x}^{*(1)}$	5	7	10	10	7	2	1	7	6	10
$y^{*(1)}$	25	49	100	100	49	4	1	49	36	100
$\mathbf{x}^{*(2)}$	8	7	6	3	10	6	1	7	6	1
$y^{*(2)}$	64	49	36	9	100	36	1	49	36	1
$\mathbf{x}^{*(3)}$	9	4	3	2	4	3	7	1	3	3
$y^{*(3)}$	81	16	9	4	16	9	49	1	9	9
$\mathbf{x}^{*(4)}$	9	5	8	7	8	2	10	9	1	5
$y^{*(4)}$	81	25	64	49	64	4	100	81	1	25
$\mathbf{x}^{*(5)}$	4	7	3	6	7	6	5	1	6	5
$y^{*(5)}$	16	49	9	36	49	36	25	1	36	25
$\mathbf{x}^{*(6)}$	2	5	5	6	9	8	9	1	3	5
$y^{*(6)}$	4	25	25	36	81	64	81	1	9	25
$\mathbf{x}^{*(7)}$	10	8	5	4	1	8	6	2	5	2
$y^{*(7)}$	100	64	25	16	1	64	36	4	25	4
$\mathbf{x}^{*(8)}$	8	4	10	1	9	7	6	7	8	1
$y^{*(8)}$	64	16	100	1	81	49	36	49	64	1
$\mathbf{x}^{*(9)}$	9	1	3	2	10	1	6	7	7	9
$y^{*(9)}$	81	1	9	4	100	1	36	49	49	81
$\mathbf{x}^{*(10)}$	1	9	6	7	3	6	8	10	5	1
$y^{*(10)}$	1	81	36	49	9	36	64	100	25	1

A method proposed by Efron and Tibshirani (1993) to improve the estimated prediction risk given by (4.15) is the following. First, the apparent error is estimated:

$$\text{Aperr} = \frac{1}{nB} \sum_{b=1}^{B} \sum_{p=1}^{n} \left[y_p^{*(b)} - g_\lambda \left(\mathbf{x}_p^{*(b)}; \hat{\mathbf{w}}_n^{*(b)} \right) \right]^2 \qquad (4.16)$$

Since each wavelet network is estimated using the bootstrapped samples $D_n^{*(b)}$ and is validated on the original sample D_n, the prediction risk $\hat{P}_\lambda$ given by (4.15) can be considered to be an out-of-sample validation. On the other hand, the apparent error can be considered to be an in-sample validation. The difference between these two measures, called the *optimism*, can be estimated by

$$\text{Opt} = \hat{P}_\lambda - \text{Aperr} \qquad (4.17)$$

Finally, the optimism is added to the training error of the original training sample D_n:

$$\tilde{P}_\lambda = L_n(\hat{\mathbf{w}}_n) + \text{Opt} \qquad (4.18)$$

The number of new samples B is usually over 30 (Aczel, 1993; Efron and Tibshirani, 1993). In our implementation 50 new samples were created. It is clear that as the number of new samples B increases, the bootstrapping method becomes more accurate but also more computationally expensive. The model selection algorithm using the bootstrapped method described above is illustrated in Figure 4.2.

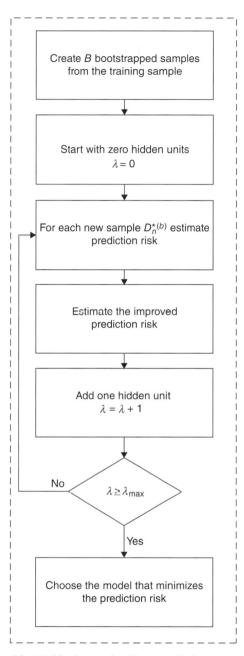

Figure 4.2 *Model selection algorithm using the bootstrap method.*

As in the case of information criteria, the prediction risk is expected to decrease monotonically until it reaches a minimum and then to increase monotonically. The number of hidden units that minimizes the prediction risk is selected for construction of the network.

Cross-Validation

Cross-validation is a standard tool for estimating the prediction error. The idea of simple validation is to split the training sample $D_n = \{x_p, y_p\}_{p=1}^n$ into two parts: the training sample $D_{\text{train}} = \{x_p, y_p\}_{p=1}^m$ and the validation sample $D_{\text{valid}} = \{x_p, y_p\}_{p=1}^{n-m}$ with $m < n$. Hence, we can train the network on the training sample and estimate the prediction risk from the new data of the validation sample. However, additional data are often not available. In simple validation not all the data available are used for network training. Hence, available information is lost and it is not utilized during the training phase of the network.

Cross-validation makes efficient use of the information available (Efron and Tibshirani, 1993). In the leave-one-out cross-validation proposed by Mosteller and Tukey (1968), the validation sample consists of only one training pattern. The procedure is the following. First, we assume that the data consist of n independently distributed random vectors. Starting with zero hidden units at step j, the jth training pair $\{x_j, y_j\}$ is removed from the training sample. Then a wavelet network is trained using the reduced sample D_{train}. The trained wavelet network, $g_\lambda(\mathbf{x}; \hat{\mathbf{w}}_{n-1}^{(j)})$, is validated on the validation sample D_{valid}, which consists of the jth training pair $\{x_j, y_j\}$.

This procedure is repeated n times and the *cross-validation criterion* is given by the equation

$$\text{CV} = \frac{1}{n} \sum_{p=1}^n \left[y_p - g_\lambda \left(\mathbf{x}_p; \hat{\mathbf{w}}_n^{(j)} \right) \right]^2 \tag{4.19}$$

and it is used as an estimator for the prediction risk $E[L(\hat{\mathbf{w}}_n)]$ for the wavelet network $g_\lambda(\mathbf{x}; \hat{\mathbf{w}}_n)$. Then 1 hidden unit is added to the network and the procedure is repeated up to a predefined maximum number of hidden units. The number of hidden units that generates the smallest prediction risk is the number of the appropriate wavelets for construction of the wavelet network. Again, it is expected that the prediction risk will decrease (almost) monotonically until it reaches a minimum and then will increase (almost) monotonically.

However, the leave-one-out cross-validation is very computationally expensive since HU · n networks must be trained. Hence, for large data sets where the training patterns are several hundreds or thousands, this method is very cumbersome and time consuming.

Alternatively, *v-fold* cross-validation can be used. In this procedure, in the first step, v new subsamples D_m^j of size $m < n$ are created with random sampling without replacement from the original training sample. Next, starting with zero hidden units, the subsamples D_m^j are removed one by one from the original sample D_n, and the network is trained on the remaining data. The resulting weight parameters are defined by the vector $\hat{\mathbf{w}}_m^{(D_j)}$. Then the trained network is validated at the left-out subsample D_m^j by estimating the mean squared cross-validation error:

$$\mathrm{CV}_{D_j} = \frac{1}{n} \sum_{\{x_p, y_p\} \in D_j} \left[y_p - g_\lambda \left(\mathbf{x}_p; \hat{\mathbf{w}}_m^{(D_j)} \right) \right]^2 \tag{4.20}$$

The prediction risk is the average mean squared cross-validation error of all subsamples and is given by

$$\hat{P}_\lambda \equiv \mathrm{CV}_\lambda = \frac{1}{v} \sum_{j=1}^{v} \mathrm{CV}_{D_j} \tag{4.21}$$

After the estimation of the prediction risk, 1 hidden unit is added to the network and the procedure is repeated up to a predefined maximum number of hidden units. The number of hidden units that produce the smallest prediction risk is the number of appropriate wavelets for construction of the wavelet network. The model selection algorithm using the cross-validation is illustrated in Figure 4.3.

To illustrate how cross-validation works, a very simple example is presented. Let us assume that the variable $x = \{1, 2, \dots, 10\}$ and that the dependent variable y is given by the relationship $y_p = x_p^2$. Hence, $y = \{1, 4, \dots, 100\}$. Our original data set consists of 10 patterns $\{x_p, y_p\}$. By applying cross-validation we will create five new samples of size 10 each. Each sample is separated in training and in a test subsample. Moreover, the training sample consists of the 80% of the observations with the validation conducted on the remaining 20%. Note that in contrast to bootstrapping the pairs (x_p, y_p) are selected randomly from the original data set without replacement. An example of five samples is presented in Table 4.2. In the case of the leave-one-out cross-validation, 10 samples would be created and the test sample would contain only one value. It is clear in this simple example that by using *v-fold* cross-validation, the computational burden of the algorithm can be reduced significantly.

Since $v \ll n$, *v-fold* cross-validation is significantly less computationally expensive than is leave-one-out cross-validation. As v, increases, the computational burden increases but also the accuracy of the method increases. When $v = n$ the leave-one-out cross-validation is retrieved. In our implementation, the training data were split into 50 subsamples.

Model Selection Without Training

The preferred information criteria were estimated by Zhang (1997) after the initialization stage of the network was performed. More precisely, in the SSO and RBS the preferred information criteria were evaluated after the selection of each wavelet in the initialization stage. Similarly, when the BE algorithm was used, the preferred information criteria were evaluated after the elimination of each wavelet in the initialization stage.

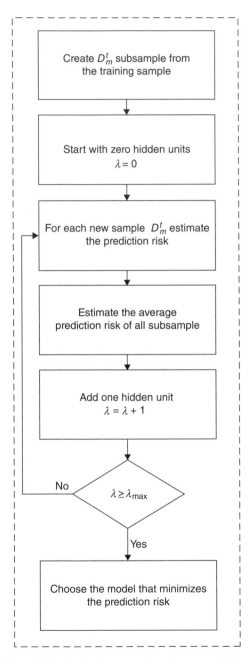

Figure 4.3 *Model selection algorithm using the cross-validation method.*

TABLE 4.2   Simple Cross-Validation Example

	Pattern								Test	Training
Sample	1	2	3	4	5	6	7	8	9	10
$x^{*(1)}$	1	9	10	5	7	2	4	3	6	8
$y^{*(1)}$	1	81	100	25	49	4	16	9	36	64
$x^{*(2)}$	2	9	8	4	10	1	3	6	7	5
$y^{*(2)}$	4	81	64	16	100	1	9	36	49	25
$x^{*(3)}$	5	8	9	10	3	4	6	7	2	1
$y^{*(3)}$	25	64	81	100	9	16	36	49	4	1
$x^{*(4)}$	9	8	10	6	1	2	7	5	3	4
$y^{*(4)}$	81	64	100	36	1	4	49	25	9	16
$x^{*(5)}$	3	1	7	6	2	8	5	4	9	10
$y^{*(5)}$	9	1	49	36	4	64	25	16	81	100

Since initialization of the wavelet network is very good, as discussed earlier, the initial approximation is expected to be very close to the target function. Hence, a good approximation of the prediction risk is expected to be obtained. The same idea can also be applied when bootstrapping or cross-validation is used.

As presented in earlier chapters, the computational burden and time needed for the initialization phase of a wavelet network are insignificant compared to the training phase. Hence, the procedure above is significantly less computationally expensive. However, the procedure is similar to early stopping techniques. Usually, early stopping techniques suggest a network with more hidden units than necessary, although the network is not fully trained to avoid overfitting (Samarasinghe, 2006). From our experience this method does not work satisfactorily in complex problems.

EVALUATING THE MODEL SELECTION ALGORITHM

To find an algorithm that will work well with wavelet networks and lead to a good estimation of prediction risk, in this section we compare the various criteria as well as the sampling techniques discussed earlier. More precisely, in this section we compare the sampling techniques that are used extensively in various studies with sigmoid neural networks and two information criteria proposed previously in the construction of a wavelet network. More precisely, the FPE proposed by Zhang (1994), the GCV proposed by Zhang (1997), the bootstrapping (BS) and *v-fold* cross-validation (CV) methods proposed by Efron and Tibshirani (1993) and Zapranis and Refenes (1999) are tested as well as the performance of the BIC criterion.

The following procedure is followed to evaluate each method. First, the prediction risk according to each method is estimated up to a predefined maximum of hidden units. Then the number of hidden units that minimizes the prediction risk is selected for construction of the wavelet network, which will be fully trained. Finally, the MSE between the wavelet network output and the target function is estimated. The best network topology will be considered the one that produces the smallest MSE and shows no signs of overfitting.

The four methods are evaluated using the functions $f(x)$ and $g(x)$ introduced in Chapter 3. Both training samples consist of 1.000 training patterns, as in the preceding section. The wavelet networks are trained using the backpropagation algorithm with 0.1 learning rate and zero momentum. To estimate the prediction risk using the bootstrapping approach, 50 new networks were created for each hidden unit ($B = 50$). Similarly, the prediction risk using the cross-validation method was estimated using 50 subsamples for each hidden unit. In other words, 50-fold cross-validation was used ($v = 50$). All wavelet networks were initialized using the BE algorithm since our results in previous sections indicate that the BE outperforms the alternative algorithms.

Case 1: Sinusoid and Noise with Decreasing Variance

In this section we focus on the function $f(x)$. As in Chapter 3, the function $f(x)$ is given by

$$f(x) = 0.5 + 0.4 \sin 2\pi x + \varepsilon_1(x) \qquad x \in [0, 1]$$

where x is equally spaced in $[0, 1]$ and the noise $\varepsilon_1(x)$ follows a normal distribution with mean zero and a decreasing variance

$$\sigma_\varepsilon^2(x) = 0.05^2 + 0.1(1 - x^2)$$

In the first case we estimate the prediction risk for a wavelet network with no hidden units, and iteratively, 1 hidden unit is added until a maximum number of 20 hidden units is reached. Table 4.3 presents the prediction risk and the suggested hidden units for each information criterion for the two functions described previously. Four of the five criteria—the FPE, BIC, BS, and CV—suggest that a wavelet network with only 2 hidden units is sufficient to model function $f(x)$. On the other hand, using the GCV, the prediction risk is minimized when a wavelet network with 3 hidden units is used.

First, we examine graphically the performance of each criterion. Figure 4.4 shows the approximation of the wavelet network to the training data using (a) 1 hidden unit, (b) 2 hidden units, and (c) 3 hidden units. Figure 4.4d shows the training data and

TABLE 4.3   Prediction Risk and Hidden Units for the Four Information Criteria[a]

	FPE	GCV	BIC	BS	CV
Case 1					
Prediction risk	0.01601	0.03149	0.03371	0.03144	0.03164
Hidden units	2	3	2	2	2
Case 2					
Prediction risk	0.00231	0.00442	0.00524	0.00490	0.03309
Hidden units	8	15	8	8	8

[a]Case 1 refers to function $f(x)$ and case 2 to function $g(x)$. FPE, final prediction error; GCV, generalized cross-validation; BS, bootstrapping; CV, 50-fold cross-validation.

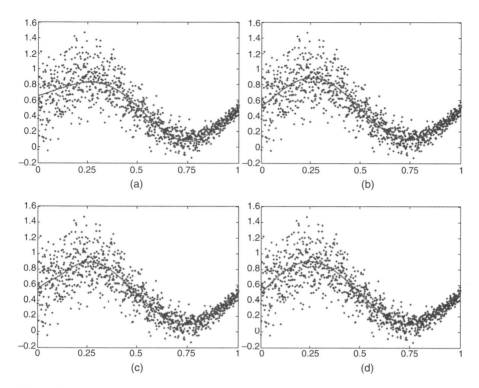

Figure 4.4 *Training a wavelet network with (a) 1, (b) 2, and (c) 3 hidden units. The target function is presented in part (d).*

the target function $f(x)$. It is clear that a wavelet network with only 1 hidden unit cannot learn the underlying function. On the other hand, wavelet networks with 2 and 3 hidden units approximate the underlying function very well. However, when 3 hidden units are used, the network approximation is affected by the large variation of the noise in the interval [0, 0.25]. To confirm the results above, the MSE between the output of the wavelet network and the underlying target function $f(x)$ is estimated. The MSE is 0.001825 when a wavelet network with only 1 hidden unit is used. Adding 1 more hidden unit, 2 in total, the MSE is reduced to only 0.000121. Finally, when 3 hidden units are used, the MSE increased to 0.000267. Hence, two wavelets should be used to construct a wavelet network to approximate function $f(x)$. The results above indicate that GCV suggested a more complex model than needed. Moreover, a wavelet network with 3 hidden units shows signs of overfitting.

From Table 4.3 it is shown that the FPE criterion suggests 2 hidden units; however, the prediction risk is only 0.01601, in contrast to GCV, BIC, BS, and CV, which is 0.03149, 0.03371, 0.03144, and 0.03164, respectively. To find the correct magnitude of the prediction risk, a validation sample is used to measure the performance of the wavelet network with 2 hidden units in out-of-sample data. The validation sample consists of 300 patterns randomly generated by $f(x)$. These patterns were not used

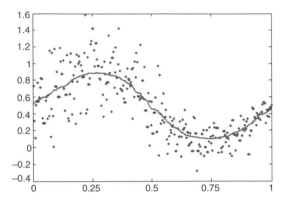

Figure 4.5   *Out-of-sample prediction for the first case.*

for training of the wavelet network. The MSE between the network forecasts and the out-of-sample targets is 0.048751, indicating that the FPE criterion is too optimistic as to estimation of the prediction risk. The approximations forecast for the wavelet network and the out-of-sample target values are shown in Figure 4.5.

Case 2: Sum of Sinusoids and Cauchy Noise

In the second part of Table 4.3, the results are presented for the model selection algorithm for the function $g(x)$. The function $g(x)$ is given by

$$g(x) = 0.5x \sin x + \cos^2 x + \varepsilon_2(x) \qquad x \in [-6, 6]$$

and $\varepsilon_2(x)$ follows a Cauchy distribution with location 0 and scale 0.05. As in the first case, the prediction risk for a wavelet network with no hidden units is estimated, and 1 hidden unit is added iteratively to the wavelet network until the predefined maximum number of 20 hidden units is reached. The FPE criterion suggests that 7 hidden units are appropriate for modeling the function $g(x)$. On the other hand, using GCV, the prediction risk is minimized when a wavelet network with 15 hidden units is used. Finally, using the BIC, bootstrapping, and the cross-validation criteria, the prediction risk is minimized when a wavelet network with 8 hidden units is used.

In Figure 4.6 the approximation of the wavelet network to the training data using (a) 7, (b) 8, and (c) 15 hidden units is presented. Part (d) of the figure shows the target function $g(x)$ and the training data. It is clear that all networks produce similar results, and it is difficult to compare them visually. To compare the results above, the MSE between the output of the wavelet network and the underlying target function $g(x)$ was estimated. The MSE is 0.001611 when a wavelet network with only 7 hidden units is used. Adding one more hidden unit, 8 in total, the MSE is reduced to only 0.000074, which is also the minimum MSE achieved. Adding additional hidden units results in an increase in the MSE between the underlying function $g(x)$ and the wavelet network. Finally, when 15 hidden units are used, the MSE increased to

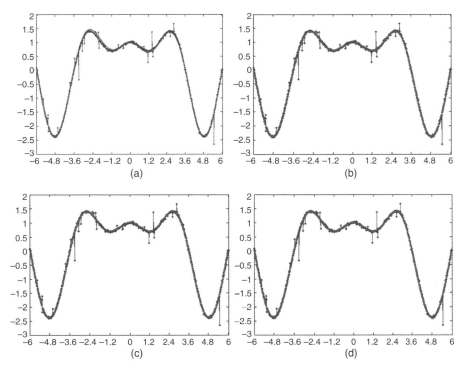

Figure 4.6   *Training a wavelet network with (a) 7, (b) 8, and (c) 15 hidden units. The target function is presented in part (d).*

0.000190. To find the best network, the MSE between the network approximation and the underlying function is estimated for a wavelet network with up to 20 hidden units. The MSE is minimized when a network with 8 hidden units is used. Hence, the optimum number of wavelets to approximate the function $g(x)$ is 8. The results above indicate that GCV suggests a more complex model, while the FPE suggests a simpler model than needed. On the other hand, our results indicate that the BIC and the sampling techniques again proposed the correct topology of the wavelet network.

As reported in Table 4.3, the estimated prediction risk proposed by the FPE criterion is 0.002312, in contrast to GCV, BIC, BS, and CV, which is 0.00442, 0.005238, 0.00490, and 0.00331, respectively. To find the correct magnitude of the prediction risk a validation sample is used to measure the performance of the wavelet network with 8 hidden units in out-of-sample data. The validation sample consists of 300 patterns randomly generated by the function $g(x)$. These patterns were not used for the training of the wavelet network. The MSE between the network forecasts and the out-of-sample targets is 0.0043. Our results again indicate that the FPE criterion is too optimistic on the estimation of the prediction risk. The out-of-sample data and the forecast produced by the wavelet network are shown in Figure 4.7.

A closer inspection of Figure 4.6 reveals that the wavelet network approximation was not affected by the presence of large outliers, in contrast to the findings of Li

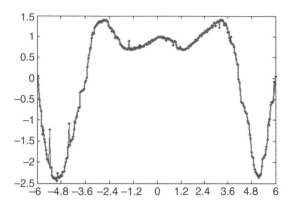

Figure 4.7   *Out-of-sample prediction for the second case.*

and Chen (2002). In this study 8 hidden units were used to construct the wavelet network, as proposed by v-fold cross-validation and bootstrapping, while in Li and Chen (2002) the architecture of the wavelet network had 10 hidden units, as proposed by the FPE criterion. Our results indicate that the FPE criterion does not perform as well as sampling techniques (bootstrapping or v-fold cross-validation) and should not be used.

Model Selection without Training   Estimation of the preferred information criteria was performed by Zhang (1997) after the initialization stage of the network. More precisely, in the SSO and RBS the preferred information criterion is evaluated after the selection of each wavelet in the initialization stage. Similarly, when the BE algorithm is used, the preferred information criteria are evaluated after the elimination of each wavelet in the initialization stage. Since initialization of the wavelet network is very good, the initial approximation is expected to be very close to the target function. Hence, a good approximation of the prediction risk is expected to be obtained. The same idea can also be applied when bootstrap or cross-validation is used. The procedure above is significantly less computationally expensive since training additional wavelet networks is not required for the bootstrapped samples.

However, the procedure above is similar to early stopping techniques. Usually, early stopping techniques suggest a network with more hidden units than necessary, although the network is not fully trained to avoid overfitting (Samarasinghe, 2006), while they do not work satisfactorily in complex problems.

In the first case the results were similar to the case where the wavelet networks were fully trained. More precisely, the FPE, bootstrap, and cross-validation methods suggested that a wavelet network with 2 hidden units is sufficient to model $f(x)$, while GCV suggested a wavelet network with 3 hidden units. In the second case, both the information criteria and sampling techniques suggested that a wavelet network with more than 14 hidden units is needed to model function $g(x)$. On the other hand, the BIC criterion suggested correctly that 8 hidden units should be used for construction of the wavelet network. However, as mentioned, the BIC assumes the asymptotic normality of the maximum likelihood estimators. Moreover, from our experience, in general, information criteria do not perform very well in complex applications. The results above indicate that when more complex problems are introduced, as in the second case, this method does not work satisfactorily.

Since sampling techniques are computationally expensive methods, the BIC criterion can be used initially to approximate the optimal number of wavelons. Then the bootstrap or cross-validation methods can be used (e.g., in ±5 hidden units), around the number of hidden units proposed by the BIC to define the best network topology.

ADAPTIVE NETWORKS AND ONLINE SYNTHESIS

In contrast to previous constructive methods, online approaches do not require that the number of wavelets be determined before the start of the training (Wong and Leung, 1998). For some applications where real-time responses of the wavelet network are crucial, online approaches, can be useful. In off-line approaches, if there is a change in the system parameters, the trained network may not be able to adapt to the change and has to be retrained. On the other hand, online training and synthesis methods allow the parameters to be updated after the presentation of each training pattern. New wavelets are added to the network when it is needed, while wavelets that do not contribute to the performance of the network anymore are removed.

Cannon and Slotine (1995) and Wong and Leung (1998) used online synthesis in the construction of wavelet networks. Similarly, Xu and Ho (1999) proposed and introduced a wavelet network for adaptive nonlinear system identification. The basis functions were selected online according to the local spatial frequency content of the approximated function. The adaptive weight updating was based on Lyapunov stability theory.

In general, an adaptive algorithm can be summarized in the following steps. First, the appropriate time that a new wavelet should be added to the network must be determined. Second, if a new wavelet is entered in the network, the optimal position in which it should be placed must be found. Third, it must be determined if a wavelet entered previously no longer contributes to the wavelet network approximation and if it should be removed. Finally, the stopping condition of the algorithm should be determined.

For the first step, the following procedure is utilized. Assume that, so far, a wavelet network with j wavelets has been constructed. After each iteration of the training algorithm the MSE between the network approximation and the target values is measured. If the MSE is stabilized after a certain point, a new wavelet is entered in the wavelet network. More precisely, the structure of the wavelet network is increased by one wavelon (hidden unit). To determine if the MSE is stabilized, the difference of the MSE between time step t and $t - 1$ can be compared to a fixed threshold:

$$\text{MSE}_t - \text{MSE}_{t-1} < \text{threshold} \tag{4.22}$$

or alternatively the ratio of the MSE between time step t and $t-1$:

$$\frac{\text{MSE}_t - \text{MSE}_{t-1}}{\text{MSE}_t} < \text{threshold} \tag{4.23}$$

In the second step, the optimal position of the new wavelet must be found. In other words, the newly entered wavelet must be initialized. At the previous stage, when j wavelets were used, the residuals between the wavelet network approximation and the target patterns could be found. Hence, the new wavelet, $j+1$, can be initialized on the residuals using the BE method.

Next, the wavelets that were entered in the model previously must be examined. The contribution of each wavelet to the total wavelet network output is examined. When the dynamics or the parameters of the underlying system were changed, there is a possibility that some of the wavelets no longer provide any useful information. To remove unnecessary wavelons, pruning methods can be employed.

The final problem that must be solved is to determine the stopping time of the algorithm. If a stopping criterion is not defined, it is possible to construct very large wavelet networks. Moreover, there is a possibility for the algorithm to be trapped in an infinite loop between steps 2 and 3. After a new wavelet is entered in the structure of the wavelet network, an information criterion such as the BIC can be estimated. If the new wavelet causes the BIC to increase, the algorithm stops.

Adaptive wavelet network can be useful in applications where dynamic systems are examined. More precisely, in cases where the parameters of the dynamic systems change, a response of the wavelet network is needed in real time. However, in problems of function approximation, the results from Wong and Leung (1998) indicate that this method is very prone to the initialization of the wavelet network. Their results indicate that the suggested topology of a particular function approximation problem varied from 4 to 10 hidden units.

CONCLUSIONS

When building a wavelet network a crucial decision that needs to be made is to choose the number of the wavelons or hidden units. The number of the wavelons also defines the architecture of the wavelet network. A network with fewer hidden units than needed would not be able to learn the underlying function; selecting more hidden units than needed will result in an overfitted model. In both cases the wavelet network cannot be used for forecasting. Moreover, the results of an analysis based on an overfitted or underfitted wavelet network are not credible or reliable.

In this chapter, various methods of selecting the optimal number of hidden units were presented and tested. More precisely, three information criteria and two resampling techniques were evaluated on two simulated cases. The five methods were used to estimate the prediction risk. In general, the resampling techniques outperformed the information criteria. In addition, both bootstrapping and cross-validation are not based on restrictive assumptions as in the case of information criteria. However, resampling schemes are computationally expensive methods.

The information criteria performed satisfactorily on the simple example. However, the results were not good on the more complex problem, with the exception of the BIC. In general, the results obtained from BIC were very stable, and this can constitute a guideline to reduce the computational burden of the resampling techniques.

Alternatively, when dynamic systems are examined and a response from the wavelet network is needed in real time, an adaptive wavelet network can prove useful. However, the wavelet forecasts of a wavelet that was constructed adaptively may be unstable.

REFERENCES

Aczel, A. D. (1993). *Complete Business Statistics*. Irwin, Homewood, IL.

Akaike, H. (1973). "Information theory and an extension of the maximum likelihood principle." In *Second International Symposium of Information Theory*, B. N. Petrov and F. Csaki, eds., Akademiai Kiado, Budapest, 267–281.

Akaike, H. (1974). "New look at statistical-model identification." *IEEE Transactions on Automatic Control*, Ac19(6), 716–723.

Anders, U., and Korn, O. (1999). "Model selection in neural networks." *Neural Networks*, 12(2), 309–323.

Cannon, M., and Slotine, J.-J. E. (1995). "Space–frequency localized basis function networks for nonlinear system estimation and control." *Neurocomputing*, 9, 293–342.

Curry, B., and Morgan, P. H. (2006). "Model selection in neural networks: some difficulties." *European Journal of Operational Research*, 170(2), 567–577.

Dimopoulos, Y., Bourret, P., and Lek, S. (1995). "Use of some sensitivity criteria for choosing networks with good generalization ability." *Neural Processing Letters*, 2(6), 1–4.

Efron, B., and Tibshirani, R. J. (1993). *An Introduction to the Bootstrap*. Chapman & Hall, New York.

Li, S. T., and Chen, S.-C. (2002). "Function approximation using robust wavelet neural networks." *Proceedings of ICTAI '02*, Washington, DC, 483–488.

Moody, J. E. (1992). "The effective number of parameters: an analysis of generalization and regularization in nonlinear learning systems." In *Advances in Neural Information Processing Systems*, J. E. Moody, S. J. Hanson, and R. P. Lippman, eds., Morgan Kaufmann, San Mateo, CA.

Mosteller, F., and Tukey, J. (1968). "Data analysis, including statistics". In *Handbook of Social Psychology*, G. Lindzey and E. Aronson, eds. Addison-Wesley, Reading, MA, Chap. 10.

Murata, N., Yoshizawa, S., and Amari, S. (1994). "Network information criterion-determining the number of hidden units for an artificial neural network model." *IEEE Transactions on Neural Networks*, 5(6), 865–872.

Reed, R. (1993). "Pruning algorithms: a survey." *IEEE Transactions on Neural Networks*, 4, 740–747.

Samarasinghe, S. (2006). *Neural Networks for Applied Sciences and Engineering*. Taylor & Francis, New York.

Wong, K.-W., and Leung, A. C.-S. (1998). "On-line successive synthesis of wavelet networks." *Neural Processing Letters*, 7, 91–100.

Xu, J., and Ho, D. W. C. (1999). "Adaptive wavelet networks for nonlinear system identification." *Proceedings of American Control Conference*, San Diego, CA.

Zapranis, A. D., and Haramis, G. (2001). "An algorithm for controlling the complexity of neural learning: the irrelevant connection elimination scheme." *Fifth Hellenic European Research on Computer Mathematics and Its Applications*, Athens, Greece.

Zapranis, A., and Refenes, A. P. (1999). *Principles of Neural Model Indentification, Selection and Adequacy: With Applications to Financial Econometrics*. Springer-Verlag, New York.

Zhang, Q. (1993). "Regressor selection and wavelet network construction." Technical Report, INRIA.

Zhang, Q. (1994). "Using Wavelet Network in Nonparametric Estimation." Technical Report, 2321, INRIA.

Zhang, Q. (1997). "Using wavelet network in nonparametric estimation." *IEEE Transactions on Neural Networks*, 8(2), 227–236.

Zhang, Q., and Benveniste, A. (1992). "Wavelet networks." *IEEE Transactions on Neural Networks*, 3(6), 889–898.

5

Variable Selection: Determining the Explanatory Variables

As mentioned in earlier chapters, the model identification procedure is divided into two parts: model selection and variable selection. In this chapter we focus on the second part of the model identification procedure, variable selection.

The use of absolutely necessary explanatory variables is known as the *principle of parsimony*, or *Occam's razor*. In real problems it is important for various reasons to determine correctly the independent variables. In most financial problems there is little theoretical guidance and information about the relationship of any explanatory variable with the dependent variable. As a result, unnecessary independent variables are included in the model, reducing its predictive power. The use of irrelevant variables is among the most common sources of specification error. Also, correctly specified models are easier to understand and to interpret. The underlying relationship is explained by only a few key variables, while all minor and random influences are attributed to the error term. Finally, including a large number of independent variables relative to sample size (an *overdetermined* model) runs the risk of a spurious fit.

To select the statistical significant and relevant variables from a group of possible explanatory variables, an approach involving the significance of statistical tests of hypotheses is followed. To do so, the relative contributions of the explanatory variables in explaining the dependent variable in the context of a specific model are estimated. Then the *significance* of a variable is assessed. This can be done by testing the null hypothesis that it is "irrelevant," either by comparing a test statistic with a known theoretical distribution with its critical value or by constructing confidence

Wavelet Neural Networks: With Applications in Financial Engineering, Chaos, and Classification,
First Edition. Antonios K. Alexandridis and Achilleas D. Zapranis.
© 2014 John Wiley & Sons, Inc. Published 2014 by John Wiley & Sons, Inc.

intervals for the relevance criterion. Our decision as to whether or not to reject the null hypothesis is based on a given significance level. The p-value, the smallest significance level for which the null hypothesis will not be refuted, imposes a ranking on the variables according to their relevance to the particular model (Zapranis and Refenes, 1999).

Evaluating the statistical significance of explanatory variables involves the following three stages:

- Defining a variable's "relevance" to the model
- Estimating the sampling variation of the relevance measure
- Testing the hypothesis that the variable is "irrelevant"

Before proceeding to the variable selection, a measure of relevance must be defined. For nonlinear models the derivative $\partial y / \partial x_j$ is not constant. As a result, new composite measures of the sensitivity of y on x_j were developed. Practitioners usually employ measures such as the average derivative and the average derivative magnitude. In the next sections, various sensitivity measures are presented. Alternatively, model-fitness sensitivity (MFS) criteria can be used. The model-fitness sensitivity criteria quantify the effect of the explanatory variable on the empirical loss and on the coefficient of determination, R^2.

Estimating sampling variability is an important part of this chapter. Without knowledge of the sampling distributions of the relevance measures, there is no way of knowing to which level the estimates available are affected by random sampling variation. In this chapter the bootstrapping and cross-validation methods are used to estimate the sampling distributions and to compute the p-values for variable significance hypothesis testing.

In this chapter we also outline a framework for hypothesis testing using nonparametric confidence intervals for the sensitivity and the MFS criteria in the context of variable selection. In this section, various methods of testing the significance of each explanatory variable are presented and tested. The purpose of this section is to find an algorithm that constantly gives stable and correct results when it is used with wavelet networks.

Once a variable is removed as being insignificant, the correctness of that action has to be evaluated. It is not uncommon, for example, to overestimate standard error in the presence of multicollinearity. A usual approach is to compare the reduced model with the full model on the basis of some performance criterion. For this purpose the prediction risk is used. Moreover, the adjusted coefficient of determination for degrees of freedom, $\bar{R}^2$, is computed, and we outline an iterative variable selection process with backward variable elimination. Finally, we use two case studies to demonstrate the framework we propose for variable selection.

EXISTING ALGORITHMS

In linear models the coefficient of an explanatory variable reflects the reactions of the dependent variable to small changes in the value of the explanatory variable.

However, the value of the coefficient does not provide any information about the significance of the corresponding explanatory variable. Hence, in linear models, to determine if a coefficient, and as a result an input variable, are significant, the t-statistics or p-values of each coefficient are examined. Applying such a method in wavelet networks is not a straightforward process since the coefficients (weights) are estimated iteratively and each variable contributes to the output of the wavelet network linearly through the direct connections and nonlinearly through the hidden units. Moreover, the finite-sample distribution of the network parameters is not known either; it must be estimated empirically, or asymptotic arguments have to be used.

Instead of removing the irrelevant variables, one can reduce the dimensionality of the input space. An effective procedure for performing this operation is *principal components analysis* (PCA). PCA has many advantages and has been used in many applications with great success (Khayamian et al., 2005). This technique has three effects on the data: It orthogonalizes the components of the input vectors (so that they are uncorrelated with each other), it orders the resulting orthogonal components (principal components) so that those with the largest variation come first, and it eliminates those components that contribute the least to variation in the data set. PCA is based on the following assumptions:

- The dimensionality of data can be reduced efficiently by linear transformation.
- Most information is contained in those directions where input data variance is maximum.

The PCA method generates a new set of variables, called *principal components*. Each principal component is a linear combination of the original variables. All the principal components are orthogonal to each other, so there is no redundant information. The principal components as a whole form an orthogonal basis for the space of the data. This approach will result in a significantly reduced set of uncorrelated variables, which will help to reduce the complexity of the network and prevent overfitting (Samarasinghe, 2006).

In applications where wavelet networks are used for the prediction of future values of a target variable, PCA can be proved very useful. On the other hand, in applications where wavelet networks are used for function approximation or sensitivity analysis, PCA can be proved to be cumbersome. The main disadvantage of PCA is that the principal components are usually a combination of all the available variables. Hence, it is often very difficult to distinguish which variable is important and which is not statistically significant. In addition, extra care must be taken when linking to the original variables the information resulting from principal components.

PCA cannot always be used since a linear transformation among the explanatory variables is not always able to reduce the dimension of the data set. Another disadvantage of PCA is the fact that the directions maximizing variance do not always maximize information. Finally, PCA is an orthogonal linear transformation, whereas the use of a wavelet network implies a nonlinear relationship between the explanatory variables and the dependent variable.

Wei et al. (2004) present a novel approach to term and variable selection. This method applies locally linear models together with orthogonal least squares to determine which of the input variables are significant. This algorithm ranks the variables and determines the amount of a system's output variance that can be explained by each term. The method assumes that nonlinearities in the system are relatively smooth. Then, local linear models are fitted in each interval. However, the number of locally linear models increases exponentially as the number of intervals for each independent variable is increased (Wei et al., 2004). Also, selecting the optimal operating regions for the locally piecewise linear models is usually computationally expensive (Wei et al., 2004).

A similar approach was presented by Wei and Billings (2007) based on feature subset selection. In feature selection an optimal or suboptimal subset of the original features is selected (Mitra et al., 2002). More precisely, Wei and Billings (2007) presented a forward orthogonal search algorithm by maximizing the overall dependency to detect the significant variables. This algorithm can produce efficient subsets with a direct link back to the underlying system. The method proposed assumes a linear relationship between sample features. However, this assumption is not always true, and the method may lead to a wider subset of explanatory variables.

SENSITIVITY CRITERIA

Alternatively, one can quantify the average effect of each input variable, x_j, on the output variable, y. The sensitivity of the wavelet network output according to small input perturbations of variable x_j can be estimated either by applying the *average derivative* (AvgD) or the *average elasticity* (AvgL), where the effect is presented as a percentage and is given by the following equations:

$$\text{AvgD}(x_j) = \frac{1}{n} \sum_{i=1}^{n} \frac{\partial \hat{y}}{\partial x_{ij}} \tag{5.1}$$

$$\text{AvgL}(x_j) = \frac{1}{n} \sum_{i=1}^{n} \frac{\partial \hat{y}}{\partial x_{ij}} \frac{x_{ij}}{\hat{y}} \tag{5.2}$$

Note that for an estimation of the average elasticity it is assumed that $\hat{y}_i \neq 0$. The average elasticity, apart from having a natural interpretation, takes into account the differences in magnitude between y and x. For example, assume that for the pairs $z_1 = \{x_1, y_1\}$ and $z_2 = \{x_2, y_2\}$ the derivatives of $\hat{y}$ with respect to x are equal. Then, in general, the same change in x (in value or percentage change) will not induce the same change in y, because of the differences in magnitude between x_1 and x_2 and y_1 and y_2, even if $dy_1/dx_1 = dy_2/dx_2$.

Although the average elasticity conveys more information, in both criteria cancellations between negative and positive values are possible. Natural extensions of the criteria above are the *average derivative magnitude* (AvgDM) and the *average elasticity magnitude* (AvgLM):

$$\text{AvgDM}(x_j) = \frac{1}{n} \sum_{i=1}^{n} \left| \frac{\partial \hat{y}}{\partial x_{ij}} \right| \tag{5.3}$$

$$\text{AvgLM}(x_j) = \frac{1}{n} \sum_{i=1}^{n} \left| \frac{\partial \hat{y}}{\partial x_{ij}} \right| \left| \frac{x_{ij}}{\hat{y}} \right| \tag{5.4}$$

Note that for estimation of the average elasticity magnitude, it is assumed that $\hat{y}_i \neq 0$. Equations (5.1) to (5.4) utilize the average derivative of the output of the wavelet network with respect to each explanatory variable. As in averaging procedures, a lot of information is lost, so additional criteria are introduced.

The *maximum* and *minimum derivative* (MaxD, MinD) or the *maximum* and *minimum derivative magnitude* (MaxDM, MinDM) give additional insight into the sensitivity of the wavelet network output to each explanatory variable. However, these criteria generally cannot be used on their own since they are appropriate only for some applications and are sensitive to inflection points (Zapranis and Refenes, 1999).

$$\text{MaxD}(x_j) = \max_{i=1...n} \left\{ \frac{\partial \hat{y}}{\partial x_{ij}} \right\} \tag{5.5}$$

$$\text{MinD}(x_j) = \min_{i=1...n} \left\{ \frac{\partial \hat{y}}{\partial x_{ij}} \right\} \tag{5.6}$$

$$\text{MaxDM}(x_j) = \max_{i=1...n} \left\{ \left| \frac{\partial \hat{y}}{\partial x_{ij}} \right| \right\} \tag{5.7}$$

$$\text{MinDM}(x_j) = \min_{i=1...n} \left\{ \left| \frac{\partial \hat{y}}{\partial x_{ij}} \right| \right\} \tag{5.8}$$

A way of "standardizing" $\partial \hat{y} / \partial x_{ij}$ is to compute the x_j's *average contribution to the magnitude of the gradient vector* since the local gradient is calculated from all derivatives (Zapranis and Refenes, 1999). Locally, the relative contribution of $\partial \hat{y} / \partial x_{ij}$ to the gradient magnitude is given by the ratio

$$r_{ij} = \frac{(\partial \hat{y}_i / \partial x_{ij})^2}{\|\nabla \hat{y}_i\|^2} = \frac{(\partial \hat{y}_i / \partial x_{ij})^2}{\sum_{i=1}^{m} (\partial \hat{y}_i / \partial x_{ij})^2} \tag{5.9}$$

and the x_j's average contribution to the gradient magnitude by

$$\text{AvgSTD}(x_j) = \frac{1}{n} \sum_{i=1}^{n} r_{ij} \tag{5.10}$$

Alternatively, the *standard deviation of the derivatives* across the sample measures the dispersion of the derivatives around their mean and is given by

$$\text{SDD}(x_j) = \left[\frac{1}{n}\sum_{j=1}^{n}\left(\frac{\partial \hat{y}}{\partial x_{ij}} - \text{AvgD}(x_j)\right)^2\right]^{1/2} \tag{5.11}$$

Finally, the standard deviation per unit of sensitivity, called the *coefficient of variation*, is given by

$$\text{CVD}(x_j) = \frac{\text{SDD}(x_j)}{\text{AvgD}(x_j)} \tag{5.12}$$

MODEL FITNESS CRITERIA

As an alternative to sensitivity criteria, model fitness criteria such as the *sensitivity-based pruning* (SBP) proposed by Moody and Utans (1992) or the effect on the *coefficient of determination* of a small pertubation of x can be used.

Sensitivity-Based Pruning   The SBP method quantifies a variable's relevance to the model by the effect on the empirical loss of the replacement of that variable by its mean. The SBP is given by

$$\text{SBP}(x_j) = L_n(\mathbf{x}; \hat{\mathbf{w}}_n) - L_n(\bar{\mathbf{x}}^{(j)}; \hat{\mathbf{w}}_n) \tag{5.13}$$

where

$$\bar{\mathbf{x}}^{(j)} = (x_{1,t}, x_{2,t}, \dots, \bar{x}_j, \dots, x_{m,t}) \tag{5.14}$$

and

$$\bar{x}_j = \frac{1}{n}\sum_{t=1}^{n} x_{j,t} \tag{5.15}$$

Additional criteria can be used, such as those presented by Dimopoulos et al. (1995).

(Adjusted) Coefficient of Determination   A model-fitness sensitivity criterion that is simpler to interpret is the effect of a variable on the sample coefficient of determination R^2, as measured by the derivative of R^2 with respect to x_j. The coefficient of determination is defined as the ratio

$$R^2 = \frac{\sum_{i=1}^{n}(\hat{y}_i - \bar{y})^2}{\sum_{i=1}^{n}(y_i - \bar{y})^2} = \frac{\text{SSR}}{\text{SST}} \tag{5.16}$$

where SSR and SST are the regression sum of squares and the total sum of squares, respectively, given by

$$SSR = \sum_{i=1}^{n} (\hat{y}_i - \bar{y})^2 \qquad (5.17)$$

$$SST = \sum_{i=1}^{n} (y_i - \bar{y})^2 \qquad (5.18)$$

The coefficient of determination can also be written as

$$R^2 = 1 - \frac{SSE}{SST} \qquad (5.19)$$

where SSE is the sum of squared residuals, given by

$$SSE = \sum_{i=1}^{n} (y_i - \hat{y}_i)^2 \qquad (5.20)$$

Note that SST is fixed for a given sample; hence, the derivative $\partial R^2 / \partial x_j$ can be computed by the following relationship:

$$\frac{\partial R^2}{\partial x_j} = \frac{1}{SST} \frac{\partial SSR}{\partial x_j} \qquad (5.21)$$

where

$$\frac{\partial SSR}{\partial x_j} = 2 \sum_{i=1}^{n} \frac{\partial \hat{y}_i}{\partial x_j} (\hat{y}_i - \bar{y}) \qquad (5.22)$$

Hence, relationship (5.21) can be written as

$$\frac{\partial R^2}{\partial x_j} = \frac{2}{SST} \sum_{i=1}^{n} \frac{\partial \hat{y}_i}{\partial x_j} (\hat{y}_i - \bar{y}) \qquad (5.23)$$

Alternatively, the adjusted coefficient of determination, $\bar{R}^2$, can be estimated. The adjusted coefficient of determination is an attempt to take into account automatically the phenomenon of R^2 and increasing spuriously when extra explanatory variables are added to the model. $\bar{R}^2$ is given by

$$\bar{R}^2 = 1 - \frac{SSE/df_e}{SST/df_t} = 1 - \frac{SSE/(n-p)}{SST/(n-1)} \qquad (5.24)$$

where df is the number of degrees of freedom, n is the number of training patterns, and p is the number of parameters that are adjusted during the training phase of the wavelet network. The number of parameters is computed by

$$p = 1 + 2\lambda j + \lambda + j = 2(\lambda + j) + j + 1 \qquad (5.25)$$

where λ is the number of hidden units and j is the number of input variables.

ALGORITHM FOR SELECTING THE SIGNIFICANT VARIABLES

To test statistically whether or not a variable is insignificant and can be removed from the training data set, the distributions of the criteria presented in the preceding section are needed. Without the distribution of the preferred measure of relevance, it is not clear if the effects of the variable x_i on y are statistically significant (Zapranis and Refenes, 1999). More precisely, the only information obtained by criteria described in the preceding section is how sensitive the dependent variable is to small perturbations of the independent variable. It is clear that the smaller the value of the preferred criterion, the less significant the corresponding variable is. However, there is no information as to whether or not this variable should be removed from the model.

We use the bootstrap method to approximate asymptotically the distribution of the measures of relevance. More precisely, a number of bootstrapped training samples can be created by the original training data set. The idea is to estimate the preferred criterion on each bootstrapped sample. If the number of the bootstrapped samples is large, a good approximation of the empirical distribution of the criterion is expected. Obtaining an approximation of the empirical distributions, confidence intervals and hypothesis tests can be constructed for the value of the criterion. The variable selection algorithm is explained analytically below and is illustrated in Figure 5.1.

The algorithm starts with a training sample that consists of all available explanatory variables.

1. Create B bootstrapped training samples from the original data set.
2. Identify the correct topology of the wavelet network following the procedure described Chapter 4 and estimate the prediction risk.
3. Estimate the preferred measure of relevance for each explanatory variable for each of the B bootstrapped training samples.
4. Calculate the p-values of the measure of relevance.
5. Test if any explanatory variables have a p-value greater than 0.1. If variables with a p-value greater than 0.1 exist, the variable with the largest p-value is removed from the training data set; else, the algorithm stops.
6. Estimate the prediction risk and the new p-values of the reduced model. If the new estimated prediction risk is smaller than the prediction risk multiplied by a

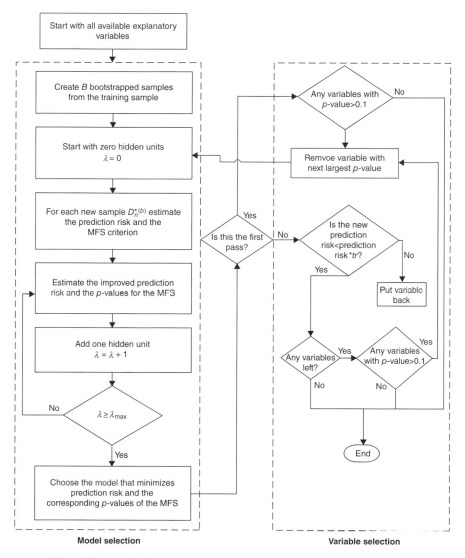

Figure 5.1 *Model identification: model selection and variable selection algorithms.*

threshold (usually, 5%), the decision of removing the variable was correct and we return to step 5.

7. If the new prediction risk is greater than the new prediction risk multiplied by a threshold (usually, 5%), the decision to remove the variable was wrong and the variable must be reintroduced to the model. In this case the variable with the next largest p-value which is also greater than 0.1 is removed from the training sample and we return to step 6. If the remaining variables have p-values smaller than 0.1, the algorithm stops.

Resampling Methods for the Estimation of Empirical Distributions

To have a good estimation of the prediction risk as well as an approximation of the distribution of the measure of relevance, a large number of bootstrapped samples B are needed. As B increases, the accuracy of the algorithm also increases, but so does the computational burden. Zapranis and Refenes (1999) presented two different bootstrap methods, *local bootstrap* and *parametric sampling*, that are significantly less computationally expensive.

Since a unique strong solution of the loss function $L(\mathbf{x}; \hat{\mathbf{w}}_n)$ does not exist, the local bootstrap is proposed by Zapranis and Refenes (1999). In local bootstrapping, first a network is trained where the original training sample is used as an input and the vector $\hat{\mathbf{w}}_n$ that minimizes the loss function is estimated. Then new samples are generated from the original training patterns using the bootstrap method. To train the new samples, the neural networks are not initialized randomly; on the contrary, the initial conditions are given by the vector $\hat{\mathbf{w}}_n$ estimated by the initial training sample. Starting the training of the neural network very close to $\hat{\mathbf{w}}_n$, the probability of the convergence of the neural network to another local minimum is reduced significantly.

A novel approach (parametric sampling) is also presented by Zapranis and Refenes (1999). The distribution of the weights of a neural network is known. As has been shown by Galland and White (1988) and White (1989), the asymptotic distribution of $\sqrt{n}(\hat{\mathbf{w}}_n - \mathbf{w}_0)$ is a multivariate normal distribution with zero mean and known covariance matrix $\mathbf{C}$, where $\hat{\mathbf{w}}_n$ is the estimated vector and $\mathbf{w}_0$ is the true vector of parameters that minimizes the loss function. Since $\mathbf{w}_0$ is not known, an estimator $\hat{\mathbf{C}}_n$ of the covariance matrix has to be used.

A basic requirement of parametric sampling is *locally identified models*, models without superfluous connections. The most important assumption is that the network was not converged in a flat minimum (i.e., $\hat{\mathbf{w}}_n$ has to be a local unique solution). This can be avoided if the irrelevant connections are removed using pruning techniques. As a result, an estimate of the standard error of any function of $\hat{\mathbf{w}}_n$ can be estimated robustly (Zapranis and Refenes, 1999).

In the framework above, instead of creating new bootstrapped training samples, Zapranis and Refenes (1999) propose sampling from the distribution of $\hat{\mathbf{w}}_n$ (parametric sampling). As a result, a very large number of parameter vectors $\hat{\mathbf{w}}_n^{(a)}$ can be created. Then any function of $\hat{\mathbf{w}}_n$ can be estimated using the bootstrapped parameter vectors $\hat{\mathbf{w}}_n^{(a)}$. This procedure can be applied to any function, like that of the measures of significance of the explanatory variables presented in the preceding section. The proposed scheme is orders of magnitude faster than alternative methods since there is no need to train new networks. For each new parameter vector the corresponding model fitness or sensitivity criterion is evaluated. However, parametric sampling is computationally more complex, since the computation and inversion of the Hessian matrix of the loss function must be estimated in order to compute $\hat{\mathbf{C}}_n$.

However, the bootstrapped samples may differ significantly from the original sample. Hence, applying local bootstrapping or parametric sampling may lead to wavelets outside their effective support (i.e., wavelets with value of zero), since wavelets are local functions with limited duration. In addition, in contrast to the case of neural networks, the asymptotic distribution of the weights of a wavelet network is not known. These observations constitute both local bootstrapping and parametric sampling inappropriate for wavelet networks.

Alternatively, new samples from training patterns can be constructed. This can be done by applying bootstrapping from pairs and training a wavelet network for each sample. Since the initialization of a wavelet network is very good, this procedure is not prohibitively expensive. The computational cost of this algorithm is B, the number of wavelet networks that must be trained. As *v-fold* cross-validation is much less computationally expensive and performs as well as bootstrapping, an approach where 50 new training samples are created according to v-fold cross-validation can also be employed.

EVALUATING THE VARIABLE SIGNIFICANCE CRITERIA

In this section the algorithm proposed in the preceding section to select the significant explanatory variables is evaluated. More precisely, the eight sensitivity criteria and the model fitness sensitivity criterion are evaluated for the two functions, $f(x)$ and $g(x)$, presented in Chapter 3.

Case 1: Sinusoid and Noise with Decreasing Variance

First, a second variable is created, which was drawn randomly from the uniform distribution within the range (0,1). Both variables are considered significant and constitute the training patterns $(\mathbf{x}_i, y_i)$ of the training data set, where $\mathbf{x}_i = \{x_{1,i}, x_{2,i}\}$ and y_i are the target values. A wavelet network is trained in order to learn the target function $f(x)$, where both x_1 and x_2 are introduced to the wavelet network as input patterns. The BE algorithm was used for initialization of the wavelet network. Using cross-validation and bootstrapping, the prediction risk is minimized when 3 hidden units are used and is 0.04194. The network converges after 3502 iterations. Comparing the results with the findings in the preceding section, it is clear that including an irrelevant variable in our model increases the model complexity and the training time while reducing the predictive power of the model. Note that as presented in Chapter 4, when only the relevant variable was used, the optimal number of hidden units was 2, while the network converged after only one iteration.

After the wavelet network is fully trained, the various measures of relevance presented in the preceding section can be estimated. The weights of the direct connections between the explanatory variables and the network output are presented in Table 5.1. In addition, the following sensitivity criteria are also presented: MaxD, the MinD, MaxDM, MinDM, AvgD, AvgDM, AvgL, and AvgLM. Finally, the SBP criterion can be found in the final column of the table. The first part of the table refers to the full model, where both variables x_1 and x_2 were used, while the second part refers to the reduced model, where only the significant variable x_1 was used.

TABLE 5.1   Sensitivity Measures for the First Case[a]

	$w_i^{[0]}$	MaxD	MinD	MaxDM	MinDM	AvgD	AvgDM	AvgL	AvgLM	SBP
Full model (two variables)										
X_1	0.0161	1.3962	−1.3459	1.3962	0.0005	−0.0529	0.6739	0.2127	1.6323	0.0953
X_2	0.0186	0.4964	−0.7590	0.7590	0.0002	0.0256	0.0915	0.0781	0.1953	0.0001
Reduced model (one variable)										
X_1	0.1296	1.1646	−1.1622	1.1644	0.0014	0.0841	0.7686	0.3165	1.3510	0.0970

[a]$w_i^{[0]}$, the linear connection between the input and output variables; MaxD, maximum derivative; MinD, minimum derivative; MaxDM, maximum derivative magnitude; MinDM, minimum derivative magnitude; AvgD, average derivative; AvgDM, average derivative magnitude; AvgL, average elasticity; AvgLM, average elasticity magnitude; SBP, sensitivity-based pruning.

Examining the direct connections of the weights between the explanatory variables and the network output, $w_i^{[0]}$, we conclude that both variables have the same significance, since both weights have almost the same value, with $w_2^{[0]}$ being slightly larger. Closer inspection of Table 5.1 reveals a contradiction between AvgL and AvdD. The AvgL value for the first variable is 0.2127 and for the second variable is 0.0781, indicating that x_1 is more significant than x_2. On the other hand, AvgD for x_2 is −0.0529 and for x_2 is 0.0256, indicating that changes in x_1 have an opposite effect on the dependent variable. In these two criteria, cancellations between negative and positive values are possible, so it is essential also to examine the AvgLM and AvgDM. Both criteria are significantly larger in the case of the first variable. The same results are obtained for the remaining sensitivity criteria. Note that MinD indicates as significant the variable with the smaller value. Finally, the SBP for the first variable is 0.0953 and for the second variable is only 0.0001, indicating that the second variable has a negligible effect on the performance of the wavelet network. Next, observing the values of the various criteria for the reduced model in Table 5.1, we conclude that the value of the weight, MinDM, and AvgD are affected by the presence of the second variable.

The previous simple case indicates that of the 10 criteria listed, only three produced robust results: AvgDM, AvgLM, and SBP. However, perhaps with the exception of SBP, it is unclear if the second variable should be removed from the training data set.

Next, the algorithm described in the preceding section will be utilized to estimate the p-values of each criterion. More precisely, the bootstrap and cross-validation methods will be applied to estimate the asymptotic distributions of the various criteria presented in Table 5.1. The mean, standard deviation, and p-values for all sensitivity and model fitness measures for the two variables of the first case are presented in Table 5.2. Using cross-validation, 50 new samples were created to approximate the empirical distributions of the various criteria, and the corresponding criteria and p-values were calculated. As expected, the average values of the criteria are similar to those presented in Table 5.1. Observing Table 5.2, it is clear that x_1

TABLE 5.2   Variable Significance Testing for the First Case Using Cross-Validation[a]

	MaxD	MinD	MaxDM	MinDM	AvgD	AvgDM	AvgL	AvgLM	SBP
Full model (two variables)									
X_1	0.9947	−1.5589	1.5589	0.0037	−0.1369	0.6898	−0.0753	1.2667	0.0967
Std.	0.0332	0.0120	0.0120	0.0023	0.0091	0.0051	0.0258	0.0375	0.0006
p-Value	0.0000	0.0000	0.0000	0.0000	0.0000	0.0000	0.0000	0.0000	0.0000
X_2	0.6231	−0.6099	0.6231	0.0001	0.0575	0.1253	0.0976	0.2137	−0.0001
Std.	0.0342	0.0182	0.0170	0.0001	0.0166	0.0031	0.0235	0.0078	0.0001
p-Value	0.0000	0.0000	0.0000	0.0000	0.0000	0.0000	0.0153	0.0000	0.6019
Reduced model (one variable)									
X_1	—	—	—	—	—	—	—	—	0.0970
Std.	—	—	—	—	—	—	—	—	0.0006
p-Value	—	—	—	—	—	—	—	—	0.0000

[a]MaxD, maximum derivative; MinD, minimum derivative; MaxDM, maximum derivative magnitude; MinDM, minimum derivative magnitude; AvgD, average derivative; AvgDM, average derivative magnitude; AvgL, average elasticity; AvgLM, average elasticity magnitude; SBP, sensitivity-based pruning.

has a larger impact on output y. However, all eight sensitivity measures consider both variables as significant predictors. As discussed previously, these criteria are application dependent; model fitness criteria are much better suited for testing the significance of the explanatory variables (Zapranis and Refenes, 1999). Indeed, the p-value for x_2 using SBP is 0.6019, indicating that this variable must be removed from the model. On the other hand, the p-value for x_1 using SBP is zero, indicating that x_1 is very significant. Finally, the p-value of x_1 using SBP in the reduced model is zero, indicating that x_1 is still very significant. Moreover, the average value of SBP is almost the same in the full and reduced models.

Next, the bootstrap method will be need to estimate the asymptotic distributions of the various criteria. To approximate the empirical distributions of the various criteria, 50 new bootstrapped samples were created, and their corresponding p-values are presented in Table 5.3. In the table the same analysis is repeated for the first case, but the random samples were created using bootstrapping. As in the cross-validation approach, 50 new bootstrapped samples were created to approximate the empirical distributions of the various criteria. A closer inspection of Table 5.3 reveals that MaxD, MinD, MaxDM, MinDM, AvgDM, and AvgLM suggest that both variables are significant and must remain in the model. On the other hand, the p-values obtained using the AvgL criterion wrongly suggests that the variable x_1 must be removed from the model. Finally, SBP and AvgD suggest correctly that x_2 must be removed from the model. More precisely, the p-values obtained using AvgD are 0.0614 and 0.3158 for x_1 and x_2, respectively, while the p-values obtained using SBP are 0 and 0.9434 for x_1 and x_2, respectively. Finally, the p-value of x_1 using SBP in the reduced model is zero, indicating that x_1 is still very significant. However, while the average value

TABLE 5.3   Variable Significance Testing for the First Case Using Bootstrapping[a]

	MaxD	MinD	MaxDM	MinDM	AvgD	AvgDM	AvgL	AvgLM	SBP
Full model (two variables)									
X_1	1.6242	−2.1524	2.2707	0.0031	−0.1079	0.6998	−0.0267	1.3498	0.0982
Std.	1.3929	2.3538	2.4426	0.0029	0.0758	0.0391	0.1651	0.4161	0.0045
p-Value	0.0000	0.0000	0.0000	0.0000	0.0614	0.0000	**0.6039**	0.0000	0.0000
X_2	1.1038	−1.2013	1.4472	0.0003	0.0402	0.1369	0.1033	0.2488	0.0011
Std.	1.4173	2.6560	2.8320	0.0003	0.0477	0.0277	0.1010	0.1244	0.0013
p-Value	0.0000	0.0000	0.0000	0.0179	0.3158	0.0000	0.4610	0.0000	0.9434
Reduced model (one variable)									
X_1	—	—	—	—	0.0800	—	—	—	0.0988
Std.	—	—	—	—	0.0433	—	—	—	0.0051
p-Value	—	—	—	—	0.0000	—	—	—	0.0000

[a]MaxD, maximum derivative; MinD, minimum derivative; MaxDM, maximum derivative magnitude; MinDM, minimum derivative magnitude; AvgD, average derivative; AvgDM, average derivative magnitude; AvgL, average elasticity; AvgLM, average elasticity magnitude; SBP, sensitivity-based pruning.

of SBP is almost the same in the full and reduced models, the average value of AvgD is completely different in magnitude and sign.

The correctness of removing a variable from the model should always be tested further. As discussed in the preceding section, this can be done either by estimating the prediction risk or the $\bar{R}^2$ of the reduced model. The prediction risk in the reduced model was reduced to 0.0396, while it was 0.0419 in the full model. Moreover, the $\bar{R}^2$ increased to 70.8% in the reduced model, while it was 69.8% in the full model. The results indicate that the decision to remove x_2 was correct.

Case 2: Sum of Sinusoids and Cauchy Noise

The same procedure is repeated for the second case, where a wavelet network is used to learn the function $g(x)$ from noisy data. First, a second variable is created which was drawn randomly from the uniform distribution within the range $(0,1)$. Both variables are considered significant and constitute the training patterns. A wavelet network is trained to learn the target function $g(x)$, where both x_1 and x_2 are introduced to the wavelet network as inputs patterns. The BE algorithm was used for initialization of the wavelet network. Using cross-validation and bootstrapping, the prediction risk is minimized when 10 hidden units are used and is 0.00336. The network approximation converges after 18,811 iterations. Again the inclusion of an irrelevant variable to our model increased the model complexity and the training time while reducing the predictive power of the model. Note that when only the relevant variable was used for the training of the wavelet network, only 8 hidden units were used, whereas the wavelet network was converged after only 1107 iterations.

TABLE 5.4   Sensitivity Measures for the Second Case[a]

	$w_i^{[0]}$	MaxD	MinD	MaxDM	MinDM	AvgD	AvgDM	AvgL	AvgLM	SBP
Full model (two variables)										
X_1	−0.0001	1.4991	−1.4965	1.4991	0.0001	0.0032	0.5517	−1.1417	6.5997	0.4202
X_2	0.0124	3.4623	−2.5508	3.4623	0.0001	0.0261	0.2691	−0.0095	0.1898	0.0002
Reduced model (one variable)										
X_1	−0.0963	1.6802	−1.5662	1.6801	0.0019	0.0011	0.6031	−0.8662	9.7935	0.4707

[a] $w_i^{[0]}$, the lineral connection between the input and output variables; MaxD, maximum derivative; MinD, minimum derivative; MaxDM, maximum derivative magnitude; MinDM, minimum derivative magnitude; AvgD, average derivative; AvgDM, average derivative magnitude; AvgL, average elasticity; AvgLM, average elasticity magnitude; SBP, sensitivity-based pruning.

Table 5.4 presents the weights of the direct connections between the explanatory variables and the network output, MaxD, MinD, MaxDM, MinDM, AvgD, AvgDM, AvgL, and AvgLM sensitivity criteria, and the SBP model fitness criterion. The first part of the table refers to the full model, where both variables x_1 and x_2 were used; the second part refers to the reduced model, where only the variable x_1 is used.

From Table 5.4 it is clear that the value of the weight of the direct connections between the first variable and the network output is smaller than the weight of the second variable. A closer inspection of the table reveals that almost all measures of relevance wrongly identify the second variable as more significant than x_1. The only exceptions are the AvgDM and AvgLM criteria. Both criteria give a significantly larger value in x_1. Finally, SBP for the first variable is 0.4202 for the first variable, while for the second variable it is only 0.0002, indicating that the second variable has a negligible effect on the performance of the wavelet network. Next, observing the values of the various criteria for the reduced model on the table, we conclude that the value of the weight, MinDM, and AvgD are affected again by the presence of the second variable.

Next, we estimate the p-values of the various criteria for the second case. The standard deviation and p-values for all sensitivity and model fitness measures for the two variables of the second case are presented in Table 5.5. Using cross-validation, 50 new samples were created to approximate the empirical distributions of the various criteria, and the corresponding criteria and p-values were calculated. The table shows that only AvgLM and SBP indentified the insignificant variable correctly. The p-values are 0.4838 and 0.6227, respectively, for the two criteria for x_2, while in the reduced model the p values of x_1 are zero. On the other hand, MinD and AvgL wrongly suggested that x_1 should be removed from the model. Finally, the remaining criteria, MaxD, MinD, MaxDM, AvgD, and AvgDM, suggest that both variables are significant and should remain in the model.

In Table 5.6 the same analysis is repeated for the second case, but the random samples were created using bootstrapping. As in the CV approach, 50 new bootstrapped

TABLE 5.5   Variable Significance Testing for the Second Case Using Cross-Validation[a]

	MaxD	MinD	MaxDM	MinDM	AvgD	AvgDM	AvgL	AvgLM	SBP
Full model (two variables)									
X_1	1.5624	−1.8437	1.8467	0.0016	−0.0192	0.5865	−2.7769	17.5475	0.4558
Std.	0.0061	0.1623	0.1561	0.0009	0.0038	0.0021	17.5304	14.7331	0.0035
p-Value	0.0000	0.0000	0.0000	**0.6897**	0.0000	0.0000	**0.7184**	0.1258	0.0000
X_2	0.7745	−1.5054	1.5054	0.0004	−0.1797	0.2349	0.0091	0.2438	0.0002
Std.	0.0214	0.0843	0.0843	0.0003	0.0056	0.0038	0.0659	0.0586	0.0002
p-Value	0.0000	0.0000	0.0000	0.0000	0.0000	0.0000	0.3539	0.4838	0.6227
Reduced model (one variable)									
X_1	—	—	—	—	—	—	—	9.4370	0.4776
Std.	—	—	—	—	—	—	—	1.8363	0.0075
p-Value	—	—	—	—	—	—	—	0.0000	0.0000

[a]MaxD, maximum derivative; MinD, minimum derivative; MaxDM, maximum derivative magnitude; MinDM, minimum derivative magnitude; AvgD, average derivative; AvgDM, average derivative magnitude; AvgL, average elasticity; AvgLM, average elasticity magnitude; SBP, sensitivity-based pruning.

samples were created to approximate the empirical distributions of the various criteria. In the table the analysis for the second case is presented. A closer inspection of the table reveals that MaxD, MinD, AvgDM, and AvgLM suggest that both variables are significant and must remain in the model. On the other hand, the p-values obtained using the AvgL and AvgD criteria wrongly suggest that the variable x_1 must be

TABLE 5.6   Variable Significance Testing for the Second Case Using Bootstrapping[a]

	MaxD	MinD	MaxDM	MinDM	AvgD	AvgDM	AvgL	AvgLM	SBP
Full model (two variables)									
X_1	1.6485	−1.8391	1.9459	0.0006	0.0225	0.5412	0.2908	8.9262	0.4191
Std.	0.3555	0.7505	0.7475	0.0008	0.0736	0.0524	7.0110	5.9525	0.0589
p-Value	0.0000	0.0000	0.0000	0.2867	**0.9877**	0.0000	**0.8708**	0.0000	0.0000
X_2	10.0490	−7.7106	11.4443	0.0007	0.0269	0.4564	−0.1217	0.6045	0.0024
Std.	16.2599	9.5366	16.9065	0.0005	0.0923	0.2912	0.5508	0.7338	0.0085
p-Value	0.07838	0.0762	0.1597	0.4158	0.6686	0.0000	0.7864	0.0000	0.8433
Reduced model (one variable)									
X_1	—	—	1.7261	0.0009	—	—	—	—	0.4779
Std.	—	—	0.0916	0.0008	—	—	—	—	0.0255
p-Value	—	—	0.0000	0.1795	—	—	—	—	0.0000

[a]MaxD, maximum derivative; MinD, minimum derivative; MaxDM, maximum derivative magnitude; MinDM, minimum derivative magnitude AvgD, average derivative; AvgDM, average derivative magnitude; AvgL, average elasticity; AvgLM, average elasticity magnitude; SBP, sensitivity-based pruning.

removed from the model. Finally, SBP, MaxD, and Min DM suggest correctly that x_2 is not a significant variable and can be removed from the model. More precisely, the p-values obtained using MaxDM are 0 and 0.1597 for x_1 and x_2, respectively, while the p-values obtained using MinDM are 0.2867 and 0.4158 for x_1 and x_2, respectively. Finally, the p-values obtained using SBP are 0 and 0.8433 for x_1 and x_2, respectively. Examining the reduced model, where only x_1 is used for the training of the WN, the p-values are zero for x_1 when the MaxDM or SBP criteria are used. On the other hand, the p-value for x_1 is 0.1795 when the MinDM is used, indicating that x_1 is insignificant and should also be removed from the model.

Next, the correctness of removing a variable from the model is tested further. As discussed in the preceding section, this can be done either by estimating the prediction risk or the $\bar{R}^2$ of the reduced model. The prediction risk in the reduced model was reduced to 0.0008, while it was 0.0033 in the full model. Moreover, the $\bar{R}^2$ increased to 99.7% in the reduced model, while it was 99.2% in the full model.

CONCLUSIONS

The criteria presented in the preceding section introduce a measure of relevance between the input and output variables. These criteria can be used for data preprocessing, sensitivity analysis, and variable selection. In this chapter we developed an algorithm for variable selection based on the empirical distribution of these criteria and examined their performance.

The results from the previous simulated experiments indicate that SBP gives constantly correct and robust results. In every case, the SBP criterion correctly identified the irrelevant variable. Moreover, the SBP criterion was stable and had the same magnitude and sign in both the full and reduced models.

The results of the previous cases indicate that when our algorithm is employed and the p-values are estimated, the performance of the remaining sensitivity criteria is unstable. In general, the sensitivity criteria were not able to identify the insignificant variable. Moreover, they often suggested removal of the significant variable x_1. The sensitivity criteria are application dependent, and extra care must be taken when they are used (Zapranis and Refenes, 1999). As their name suggests, they are more appropriate for use in sensitivity analysis than in variable significance testing.

Finally, when the bootstrap method was used, the standard deviation of each criterion was constantly significantly larger than the values obtained when the cross-validation was used. Bootstrapped samples contain more variability and may differ significantly from the original sample. As a result, an unbiased empirical distribution of the corresponding statistic is obtained.

REFERENCES

Dimopoulos, Y., Bourret, P., and Lek, S. (1995). "Use of some sensitivity criteria for choosing networks with good generalization ability." *Neural Processing Letters*, 2(6), 1–4.

Galland, A. R., and White, H. (1988). *A Unified Theory of Estimation and Inference for Nonlinear Dynamic Models*. Basil Blackwell, Oxford.

Khayamian, T., Ensafi, A. A., Tabaraki, R., and Esteki, M. (2005). "Principal component-wavelet networks as a new multivariate calibration model." *Analytical Letters*, 38(9), 1447–1489.

Mitra, P., Murthy, C. A., and Pal, S. K. (2002). "Unsupervised feature selection using feature similarity." *IEEE Transactions on Pattern Analysis and Machine Intelligence*, 24(3), 301–312.

Moody, J. E., and Utans, J. (1992). "Principled architecture selection for neural networks: application to corporate bond rating prediction." In *Neural Networks in the Capital Markets*, A. P. Refenes, ed. Wiley, New York.

Samarasinghe, S. (2006). *Neural Networks for Applied Sciences and Engineering*. Taylor & Francis, New York.

Wei, H.-L., and Billings, S. A. (2007). "Feature subset selection and ranking for data dimensionality reduction." *IEEE Transactions on Pattern Analysis and Machine Intelligence*, 29(1), 162–166.

Wei, H.-L., Billings, S. A., and Liu, J. (2004). "Term and variable selection for non-linear system identification." *International Journal of Control*, 77(1), 86–110.

White, H. (1989). "An additional hidden unit test for neglected nonlineartiy in multilayer feedforward networks." *IJCNN*, Washington, DC, 451–455.

Zapranis, A., and Refenes, A. P. (1999). *Principles of Neural Model Indentification, Selection and Adequacy: With Applications to Financial Econometrics*. Springer-Verlag, New York.

6

Model Adequacy: Determining a Network's Future Performance

In this chapter we present various metrics in order to assess a trained network. We are interested in measuring the predictive ability of a wavelet network in the context of a particular application. The evaluation of the model usually includes two clearly distinct, although related stages.

In the first stage, various metrics that quantify the accuracy of the predictions or the classifications made by the model are used and the model is evaluated based on these metrics. The term *accuracy* is a quantification of the "proximity" between the outputs of the wavelet network and the target values desired. The measurements of the precision are related to the error function that is minimized (or in some cases, the profit function that is maximized) during model specification of the wavelet network model. When an estimate of the model is done by minimizing the squared error function, the simplest example of such a measurement of accuracy is the mean squared error (MSE). The most common error criteria are the MSE, the root MSE, the normalized MSE, the sum of squared errors, the maximum absolute error, the mean absolute error, the (symmetric) mean absolute percentage error, and Theil's U index.

In addition, useful and immediate information can be provided by visual examination of a scatter plot of the network forecasts and the target values. Moreover, the statistical hypothesis testing of the values of the intercept and the slope of the regression between the wavelet network outputs and the target values can also provide useful information. Indicators such as the prediction of change in direction, independent prediction of change in direction, and the position of sign evaluate the ability of a wavelet network to predict changes in sign or direction of the values. In

Wavelet Neural Networks: With Applications in Financial Engineering, Chaos, and Classification,
First Edition. Antonios K. Alexandridis and Achilleas D. Zapranis.
© 2014 John Wiley & Sons, Inc. Published 2014 by John Wiley & Sons, Inc.

classification applications, metrics such as the classification rates, relative entropy, and the Kolmogorov–Smirnov statistic are used. The metrics above provide useful information about the predictive capabilities of the wavelet network; however, they are not sufficient for a complete evaluation of the adequacy of our model.

The second step is to assess the behavior and the performance of the wavelet network model under conditions as close as possible to the actual operating conditions of the application. The greater accuracy of the network model does not necessarily mean that it will be applied more successfully. For example, in a time-series application for forecasting returns, it is possible for a model that is characterized by low MSE to create large losses because of only a few failed predictions that are accompanied by high costs (Yao et al., 1999).

It is important, therefore, that the performance of the model is evaluated in the context of decision making that it supports. Especially in the case of time-series forecasting applications, the performance and the evaluation of the model should be based on benchmark trading strategies. It is also important to evaluate the behavior of the model throughout the range of the actual possible scenarios. For example, a trader should know how a predictive model behaves during an upward or downward trend in the market, sideways movements, if it is able to predict turning points, and so on.

The wavelet neural model is evaluated using validation samples. The patterns used in the validation samples were not used during the training of the wavelet network and represent all possible scenarios that can arise in reality. For example, if the application concerns prediction of the performance of a stock index, it would not be correct if the validation sample corresponds solely to a period with a strong upward trend since the evaluation will be restricted to the specific circumstances. It is possible to use more than one validation sample, each corresponding to different conditions, to cover the full range of possible scenarios.

The full understanding of the behavior of the model under different conditions is a prerequisite for the creation of a successful decision support system or simply a decision-making system (e.g., automated trading systems). The sensitivity analysis of the model will help us understand the dynamics and the nature of the process that the wavelet network has learned during the training phase. An optical analysis of the relationship between the explanatory and the dependent variable is recommended, by constructing two- or three-dimensional plots that connect the value of the dependent variable with the value of one or two explanatory variables, respectively. Using the model adequacy process and the appealing properties of wavelet networks, the wavelet neural model ceases to be a "black box" but, rather, it constitutes a reliable framework where we can base our decisions depending on the application. Almost always, a pilot phase follows where the final evaluation of our model is performed under the actual conditions.

TESTING THE RESIDUALS

For a "correctly specified" wavelet neural network model, the nonparametric residuals

$$y_i - g(\mathbf{x}; \hat{\mathbf{w}}_n) = e_i \qquad (6.1)$$

are such that $e_i \cong \varepsilon_i = y_i - \varphi(x_i)$. The residuals $\{e_i\}$ can be used to perform meaningful diagnostic tests about the initial assumptions regarding error term ε of the theoretical underlying model $y_i = \varphi(x_i) + \varepsilon_i$. However, because of the nonparametric nature of wavelet neural networks, satisfying those tests is a "necessary but not sufficient" condition for model adequacy. As in the case of the methodology proposed by Box and Jenkins (1970) for ARMA models, the stage of diagnostic checking should be integrated in the process of specifying a model, but it cannot replace it.

The graphical representation and visual inspection of the residuals can reveal extreme values, autocorrelations, or periodicities. To test for autocorrelation, the true nature of the serially correlated error process must be known. However, often this is not the case. As a result, the methods proposed include the fitting of stationary time series to ordinary least squares residuals. An example is the Durbin–Watson test, which has been developed for linear regression models to test the hypothesis that $\varphi = 0$ in the error model $e_i = \varphi e_{i-1} + a_i$. For linear models it has been shown that the autocorrelations estimated from the OLS residuals converge to the true autocorrelations. This observation can lead to tests such as the Box–Pierce and Ljung–Box, which are used for adequacy testing of ARMA models. These tests will still hold asymptotically for nonlinear models such as wavelet neural networks, although larger samples will be needed for finite sample validity (Zapranis and Refenes, 1999). It should be noted here that the relevant theory behind these tests refers to an observed time series. Here we use $\{e_i\}$ and we assume that it is sufficiently close to $\{\varepsilon_i\}$. In the following paragraphs we refer to various diagnostic tests, including a test for the adequacy of the model in the sense of heteroscedasticity presence in the error term. More sophisticated tests to identify the source of the specification error also exist. One typical example is Ramsey's RESET test (Ramsey, 1969), which is based on the residuals of estimated linear regression models, and it is used to identify specification errors due to omitted variables or incorrect functional form. A similar test for neural networks was proposed by White (1989).

Testing for Serial Correlation in the Residuals

Correlogram   The autocorrelation function (ACF) can be used to determine whether there is any pattern in the residuals, and the absence of such a pattern may reveal that the particular sample of residuals is random. The sample autocorrelation function for lag k, denoted by $\hat{r}_k$, is defined as follows:

$$\hat{r}_k = \frac{\hat{\gamma}_k}{\hat{\gamma}_0} \qquad (6.2)$$

where γ_k is the sample covariance (or autocovariance) at lag k and γ_0 is the sample variance (i.e., the covariance at lag 0), which are given by the following two equations:

$$\hat{\gamma}_k = \frac{\sum (e_i - \bar{e})(e_{i+k} - \bar{e})}{n} \qquad (6.3)$$

$$\hat{\gamma}_0 = \frac{\sum (e_i - \bar{e})^2}{n} \qquad (6.4)$$

where $\bar{e}$ is the mean value of the residuals and n is the sample size. A plot of $\hat{r}_k$ against k is known as the sample correlogram. It can be shown that if a time series is purely random, the sample autocorrelation coefficients are asymptotically normally distributed with zero mean and variance $1/n$. Thus, it can be concluded that the residuals are random if the autocorrelations calculated are within the limits

$$-z_a \frac{1}{\sqrt{n}} \leq \hat{r}_k \leq +z_a \frac{1}{\sqrt{n}} \tag{6.5}$$

with z_a being the $100(1-a)$ percentile point of the standard normal distribution.

Box–Pierce Q-Statistic   To test the joint hypothesis that all the autocorrelation coefficients are simultaneously equal to zero (i.e., to test the null hypothesis H_0: $r_1 = r_2 = \cdots = r_m = 0$ against the alternative that not all autocorrelation coefficients are zero), one can use the Box–Pierce test (Makridakis et al., 2008). The Box–Pierce test requires computation of the "portmanteau" Q-statistic:

$$Q = n \sum_{k=1}^{m} \hat{r}_k^2 \tag{6.6}$$

where m is the lag length. Under H_0, Q is distributed asymptotically as χ_m^2.

Ljung–Box LB-Statistic   A variant of the Box–Pierce Q-statistic is the Ljung–Box LB-statistic, defined as

$$\text{LB} = n(n+2) \sum_{k=1}^{m} \frac{\hat{r}_k^2}{n-k} \tag{6.7}$$

which under the null hypothesis is also distributed as χ_m^2. The Ljung–Box test has better properties than the Box–Pierce test. Typically, the Q-statistic is evaluated for several choices of m. Under H_0, for large n,

$$E[Q(m_1) - Q(m_2)] = m_2 - m_1 \tag{6.8}$$

so that different sections of the correlogram can be checked for departures from H_0. Furthermore, acceptance of H_0 requires one to check the individual autocorrelation coefficients to see whether a large number of them are close to zero and mask the presence of a highly significant individual.

Durbin–Watson Statistic   The Durbin–Watson test was derived for linear regression models to test the hypothesis that $\varphi = 0$ in the error model $e_i = \varphi e_{i-1} + a_i$. The test statistic is

$$d = \frac{\sum_{i=2}^{n} (e_i - e_{i-1})^2}{\sum_{i=2}^{n} e_i^2} \approx 2(1 - \hat{r}_1) \tag{6.9}$$

where $\hat{r}_1$ is the sample autocorrelation at lag $k = 1$. Durbin and Watson (1951) calculated bound d_L and d_U for d, which depend on the number of linear regressors. The null hypothesis $H_0: \varphi = 0$ versus the alternative $H_1: \varphi > 0$ is not rejected if $d > d_U$, and it is rejected if $d < d_L$. The test is inconclusive if $d_L < d < d_U$. To test versus $H_1: \varphi < 0$, we replace d with $4 - d$. The test was described analytically by Judge et al. (1982). The asymptotic equivalence (locally) of a nonlinear regression [e.g., the wavelet network model $g(\mathbf{x}; \hat{\mathbf{w}}_n)$] and its linear approximation around the true parameter vector $\mathbf{w}_0$, as given by the first-order Taylor approximation

$$g(\mathbf{x}; \mathbf{w}_0) + \frac{\partial(\mathbf{x}; \mathbf{w}_0)}{\partial \hat{\mathbf{w}}_n^T}(\hat{\mathbf{w}}_n - \mathbf{w}_0) \tag{6.10}$$

suggests that the test will be approximately valid. However, the applicability of the Durbin–Watson test for nonlinear models has not been established rigorously (Zapranis and Refenes, 1999).

EVALUATION CRITERIA FOR THE PREDICTION ABILITY OF THE WAVELET NETWORK

To evaluate the ability of the wavelet network to predict the level of values, prices, returns, or trends, various metrics are used. Generally, we can distinguish between two groups of measurements: (1) measurements regarding the accuracy of the prediction of our model in isolation or compared against another reference model, and (2) measurements regarding the predictability of the changes in direction of the values. Also, very often, visual inspection of the scatter plot between the wavelet network predictions against target values is used, as it provides immediate information. Moreover, in this section we present and analyze the predictive power of the wavelet network based on statistical hypothesis testing of the values of the parameters of the linear regression between the outputs of the wavelet network and the target values.

Measuring the Accuracy of the Predictions

The usual metric that is used for quantification of the accuracy of the predictions of the network is the mean squared error (MSE). To find the MSE, first the difference between the network predictions and the real target values is estimated. This difference is squared and then the average is taken. The MSE is given by

$$\text{MSE} = \frac{1}{n} \sum_{i=1}^{n} (y_i - \hat{y}_i)^2 \tag{6.11}$$

where y_i are the target values, $\hat{y}_i$ are the wavelet networks prediction, and n is the number of observations.

Another measure that is often used is the root mean squared error (RMSE). The RMSE is simply the squared root of the MSE. Since MSE is considered the variance of an unbiased estimator, in the same sense, the RMSE is the standard deviation. The RMSE is given by

$$\text{RMSE} = \sqrt{\frac{1}{n} \sum_{i=1}^{n} (y_i - \hat{y}_i)^2} = \sqrt{\text{MSE}} \tag{6.12}$$

An extension of the previous metric is the normalized mean squared error (NMSE), which is given by

$$\text{NMSE} = \frac{\sum_{i=1}^{n} (y_i - \hat{y}_i)^2}{\sum_{i=1}^{n} (y_i - \bar{y})^2} \tag{6.13}$$

where $\bar{y}$ is the average value of the target values:

$$\bar{y} = \frac{1}{n} \sum_{i=1}^{n} y_i \tag{6.14}$$

On the other hand, the mean absolute error (MAE) measures the accuracy in absolute terms:

$$\text{MAE} = \frac{1}{n} \sum_{i=1}^{n} |y_i - \hat{y}_i| \tag{6.15}$$

A metric similar to the MAE is the median absolute error (Md.AE). The median is the numerical value that separates the higher half of the errors from the lower half. Hence, to estimate the Md.AE, we first arrange all the absolute errors from the lowest value to the highest, and then we pick the middle one. If there is an even number of observations, there is no single middle value; in this case the Md.AE is defined to be the mean of the two middle values.

Sometimes the sum of squared errors (SSE) is used. The SSE is given by

$$\text{SSE} = \sum_{i=1}^{n} (y_i - \hat{y}_i)^2 \tag{6.16}$$

In some cases it is useful to estimate the maximum absolute error (MaxAE):

$$\text{MaxAE} = \max |y_i - \hat{y}_i| \tag{6.17}$$

Similarly, the mean absolute percentage error (MAPE) measures the accuracy of the network prediction in absolute and percentage terms:

$$\text{MAPE} = \frac{100}{n} \sum_{i=1}^{n} \left| \frac{y_i - \hat{y}_i}{y_i} \right| \tag{6.18}$$

The difference between the actual value and the prediction is again divided by the actual value. However, the concept of MAPE has a major drawback in practical applications. If there are zero values in the data sample, there will be a division by zero. In the case of a perfect fit, MAPE is zero.

The symmetrical mean absolute percentage error (SMAPE) is an extension of the MAPE. The SMAPE is given by

$$\text{SMAPE} = \frac{100}{n} \sum_{i=1}^{n} \frac{|y_i - \hat{y}_i|}{(y_i + \hat{y}_i)/2} \tag{6.19}$$

This formula provides a result between 0 and 200%. However, a percentage error between 0 and 100% is much easier to interpret. Hence, the following formula is used in practice:

$$\text{SMAPE} = \frac{100}{n} \sum_{i=1}^{n} \frac{|y_i - \hat{y}_i|}{y_i + \hat{y}_i} \tag{6.20}$$

Often, we want to compare the performance between two models. To compare the predictive power of a wavelet network and a benchmark, Theil's U Index is used:

$$U = \sqrt{\sum_{i=1}^{n} \left(\frac{y_i - \hat{y}_i}{y_i - f_i} \right)^2} \tag{6.21}$$

where f_i are the predictions of the benchmark model. If $U = 1$, the prediction power of the wavelet network is equal to the predictive power of the benchmark model. If $U > 1$, the performance of the networks is worse than the benchmark model, while if $U < 1$, the network's performance is better. The U index is generally used in forecasting econometric time series, and the benchmark model is the simple random walk.

Scatter Plots

The visual examination of a scatter plot between the forecasts of the wavelet network and the target values can provide direct information about the predictive ability of the network. In a scatter plot, the horizontal axis represents the predictions of the wavelet network, $\hat{y}_i$, and the vertical axis represents the target values, y_i. Each point on the graph corresponds to a pair of values $(y_i, \hat{y}_i)$. Clearly, it is desirable that the points are to be distributed near and around the straight line $y_i = \hat{y}_i$ which has a slope of 45° and passes through the origin of the axes. In Figures 6.1a and 6.2 we see this case exactly. On the other hand, Figure 6.1b shows a model that generates predictions rather randomly. An intermediate situation between the two corresponds to an intermediate predictor. Other possible cases are presented in Figure 6.1c and d. In the first case we have a linear relationship but with a slope different from 45°; in the second case the network has no predictive power and almost always returns the same output value.

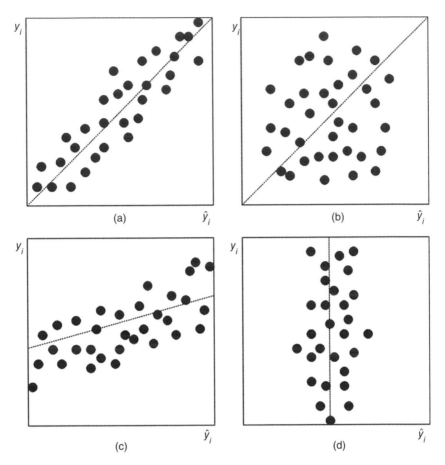

Figure 6.1   *Various scatter plots between the wavelet network's predictions and the target values.*

Linear Regression Between Forecasts and Targets

Although scatter plots provide an easy and direct (visual) assessment of the predictive power of the wavelet network, which do not quantify this ability. A very useful way to quantify the forecasting ability of a wavelet network, which also allows comparison of different models, is the linear regression between the forecasts of the wavelet network and the target values: in other words, estimation of the following model:

$$y_i = \beta_0 + \beta_1 \hat{y}_i \qquad (6.22)$$

The estimated parameters that arise from the OLS of the parameters β_0 and β_1 are denoted by b_0 and b_1, respectively. The residuals of the regression are given by $e_i = y_1 - (b_0 + b_1 \hat{y}_i)$. Hence, the estimated model is

$$y_1 = b_0 + b_1 \hat{y}_i + e_i \qquad (6.23)$$

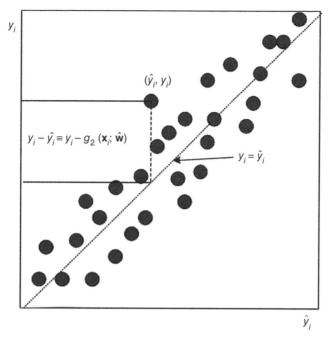

Figure 6.2 *Scatter plot between the wavelet network's predictions and the target values where the line* $y_i = \hat{y}_i$ *passes through the origin and has a slope of* 45°.

To estimate the t-statistics and p-values of the parameters b_0 and b_1, a normal distribution is assumed.

If the slope of the regression b_1 is equal to zero, there is no linear relationship between the targets y_i and the predicted values $\hat{y}_i$. There are two cases where the slope of the regression can be zero:

- When the predicted value, $\hat{y}_i$, is constant for every target value.
- When the predicted values, $\hat{y}_i$, and the targets, y_i, are completely uncorrelated.

In any other case there is a form of linear relationship. The coefficient of determination R^2 indicates how well data points fit a line or curve. The values of R^2 are between 0 and 1 and given by the following relationship:

$$R^2 = 1 - \frac{\text{SSE}_{\text{TP}}}{\text{SST}_{\text{TP}}} \tag{6.24}$$

where SSE_{TP} is the sum of the squared errors given by

$$\text{SSE}_{\text{TP}} = \sum_{i=1}^{n} [y_i - (b_0 + b_1\hat{y}_i)]^2 = \sum_{i=1}^{n} e_i^2 \tag{6.25}$$

and SST_{TP} is the total sum of squares:

$$\text{SST}_{TP} = \sum_{i=1}^{n} (y_i - \bar{y})^2 \tag{6.26}$$

We use the subscript TP to refer to the regression between the target values and the values predicted.

The F-statistic informs us if there is a linear relationship between the target values and the values predicted:

$$F = \frac{\text{SST}_{TP} - \text{SSE}_{TP}}{\text{SSE}_{TP}/(n-2)} \tag{6.27}$$

The same information can be obtained from the t-statistic of the slope of the regression b_1. More precisely, the t-statistic is used for the following hypothesis test regarding the existence of a linear relationship between the target values, y_i, and the values predicted, $\hat{y}_i$:

$$\begin{aligned} H_0: b_1 &= 0 \\ H_1: b_1 &\neq 0 \end{aligned} \tag{6.28}$$

If the slope of the linear regression is zero, there is no linear relationship between the targets y_i and the predictions $\hat{y}_i$. If it is not equal to zero, the slope will be either positive or negative, and as a result, a linear relationship will exist.

Assuming that the residuals e_i follow a normal distribution, the t-statistic follows the t-Student distribution with $v = n - 2$ degrees of freedom and is given by

$$t_{n-2} = \frac{b_1}{\text{s.e.}(b_1)} \tag{6.29}$$

where s.e. (b_1) is the standard error of the slope b_1, which is given by

$$\text{s.e.}(b_1) = \frac{s}{\sqrt{\sum_{i=1}^{n} (\hat{y}_i - \hat{\bar{y}}_i)}} \tag{6.30}$$

where s is the standard error of the residuals:

$$s = \sqrt{\frac{\sum_{i=1}^{n} e_i^2}{n-2}} = \sqrt{\frac{\text{SSE}_{TP}}{n-2}} \tag{6.31}$$

and

$$\hat{\bar{y}} = \frac{1}{n} \sum_{i=1}^{n} \hat{y}_i \tag{6.32}$$

For a large sample and a significance level $\alpha = 0.05$, the critical values of the t-Student distribution are ± 1.96. If the t-statistic is larger than 1.96 or smaller than -1.96, the hypothesis H_0 is rejected. In other words, b_1 is statistically significantly different from zero; hence, a linear relationship exists between the targets y_i and the predictions $\hat{y}_i$. Alternatively, the p-values can be calculated. The p-values report the minimum significance level at which the null hypothesis can be rejected.

The previous test is not sufficient for our analysis. Even if a linear relationship exists between the targets y_i and the predictions $\hat{y}_i$, it is possible that the imaginary line formed by the pairs $(\hat{y}_i, y_i)$ does not pass through the origin of the axes or have a slope different from 45°, or both.

When the imaginary straight line does not pass through the origin, the constant term of the regression b_0 is different from zero. Therefore, in this case the null hypothesis H_0 that the constant term is zero should be tested against the alternative H_1 that it is different from zero:

$$\begin{aligned} H_0&: b_0 = 0 \\ H_1&: b_0 \neq 0 \end{aligned} \tag{6.33}$$

The test of the hypothesis of a zero value of b_0 is based on the t-ratio

$$t_{n-2} = \frac{b_0}{\text{s.e.}(b_0)} \tag{6.34}$$

where s.e.(b_0) is the standard error of the constant b_0, which is given by

$$\text{s.e.}(b_1) = s\sqrt{\frac{1}{n} + \frac{\hat{\bar{y}}_i^2}{\sum_{i=1}^{n}(\hat{y}_i - \hat{\bar{y}}_i)^2}} \tag{6.35}$$

The t-ratio given by equation (6.34) is compared against the critical value of the t-Student distribution with $v = n - 2$ degrees of freedom.

Finally, we must test whether or not the slope of the imaginary straight line formed by the pairs $(\hat{y}_i, y_i)$ is different from 45°. In other words, we want to test the null hypothesis H_0, that the slope of the regression $b_1 = 1$, against H_1, that $b_1 \neq 1$:

$$\begin{aligned} H_0&: b_1 = 1 \\ H_1&: b_1 \neq 1 \end{aligned} \tag{6.36}$$

The t-ratio given by (6.29) is a special version of the statistic

$$t_{n-2} = \frac{b_1 - \beta_{1,0}}{\text{s.e.}(b_1)} \tag{6.37}$$

where $\beta_{1,0}$ is the value of the slope b_1 under the null hypothesis. Hence, for $b_1 = 1$ we have that

$$t_{n-2} = \frac{b_1 - 1}{\text{s.e.}(b_1)} \tag{6.38}$$

It should be noted that if autocorrelation exists in the residuals of the regression e_i, the standard errors of the intercept b_0 and the slope b_1 of the regression will be very small. As a result, it will be difficult to accept the null hypotheses H_0: $b_0 = 0$ and H_0: $b_1 = 1$. For this reason, the statistical Durbin–Watson (DW), which is used for the diagnosis of autocorrelation in the residues of regression, is also reported.

Summing up, for a properly identified, nonbiased wavelet neural network, a linear relationship between the targets and the predictions of the network should exist, the intercept b_0 should be equal to zero, and the slope of the regression b_1 should be equal to 1. It should be clarified that this is a necessary but not a sufficient condition for a proper model identification framework. For example, an overparameterized wavelet neural network model that "learned" the noise that exists in the training data, but not the underlying function that generated the observations, will satisfy the foregoing conditions for the training sample but will be biased with reduced generalization ability to new data.

The generating process of the observations is given by the relationship

$$y_i = \varphi(\mathbf{x}_i) + \varepsilon_i. \tag{6.39}$$

where the error ε_i is distributed identically and independently with mean zero and variance σ_ε^2. We have that

$$\begin{aligned} \text{var}\left[(y_i - \hat{y}_i)^2\right] &= \text{var}\left[(\varphi(\mathbf{x}_i) - \hat{y}_i)^2\right] + \text{var}\left[\varepsilon_i^2\right] \\ &\Leftrightarrow \sigma_p^2(\mathbf{x}) = \sigma_m^2(\mathbf{x}) + \sigma_e^2 \end{aligned} \tag{6.40}$$

where σ_p^2 is the prediction variance and σ_m^2 is the model variance.

As can be seen from equation (6.40) the size of the dispersion of the pairs $(\hat{y}_i, y_i)$ around the imaginary straight $y_i = \hat{y}_i$ depends on two factors: (1) how well the wavelet neural model was identified [i.e., how close the predictions of the network $\hat{y}_i$ are to the unknown underlying function $\varphi(\mathbf{x}_i)$] and (2) the variance of the error term σ_ε^2. It follows that the pairs $(\hat{y}_i, y_i)$ will lie on the imaginary straight line $y_i = \hat{y}_i$ if the model has been specified properly and there is no error term.

Measuring the Ability to Predict the Change in Direction

While the indicators discussed previously examine the accuracy of the prediction, in some cases we are also interested in predicting the changes in direction of the dependent value independent of their range. These metrics are often used in the analysis and forecasting of financial time series and are expressed as percentages.

The first indicator is the prediction of change in direction (POCID):

$$POCID = \frac{100}{n} \sum_{i=1}^{n} d_i \qquad (6.41)$$

where

$$d_i = \begin{cases} 1 & (y_i - y_{i-1})(\hat{y}_i - y_{i-1}) > 0 \\ 0 & (y_i - y_{i-1})(\hat{y}_i - y_{i-1}) \le 0 \end{cases} \qquad (6.42)$$

An extension of the POCID is the independent prediction of change in direction (IPOCID), given by

$$IPOCID = \frac{100}{n} \sum_{i=1}^{n} d_i \qquad (6.43)$$

where

$$d_i = \begin{cases} 1 & (y_i - y_{i-1})(\hat{y}_i - \hat{y}_{i-1}) > 0 \\ 0 & (y_i - y_{i-1})(\hat{y}_i - \hat{y}_{i-1}) \le 0 \end{cases} \qquad (6.44)$$

Finally, an indicator often used in forecasting financial returns is the prediction of sign (POS), given by

$$POS = \frac{100}{n} \sum_{i=1}^{n} d_i \qquad (6.45)$$

where

$$d_i = \begin{cases} 1 & y_i \cdot \hat{y}_i > 0 \\ 0 & y_i \cdot \hat{y}_i \le 0 \end{cases} \qquad (6.46)$$

TWO SIMULATED CASES

In this section we demonstrate how the previous metrics can be used in two simulated cases to assess the model adequacy of a wavelet network. The first case is the sinusoid with a decreasing variance in the noise, and the second case is a summation of two sinusoids with Cauchy noise. Both cases were discussed in earlier chapters.

Case 1: Sinusoid and Noise with Decreasing Variance

As discussed in Chapters 3 and 4, a wavelet network with 2 hidden units was used. The initialization method was presented in Chapter 3 and various methods for selecting the optimal architecture of the wavelet network were presented in Chapter 4. In this

TABLE 6.1   Case 1: Residuals Testing[a]

	Parameter	p-Values
n/p Ratio	125.1250	
Mean	0.0001	
Median	0.0045	
S. dev.	0.1771	
DW	1.9594	0.5199
LB Q-stat.	42.5310	0.0024
JB stat.	74.26664	0.0010
KS stat.	10.5471	0.0000
R^2	0.71044	
$\bar{R}^2$	0.70840	

[a]S. dev., standard deviation; DW, Durbin–Watson; LB, Ljung–Box; KS, Kolmogorov–Smirnov.

section we examine how well the wavelet network was trained on the data: in other words, the ability of the network to learn the data and then forecast the future values of the underlying function.

Testing the Residuals   First, the residuals of the trained networks will be examined. In Table 6.1 the various descriptive statistics as well as various tests on the residuals are presented. The mean of the residuals is close to zero with a standard deviation of 0.17. Both the Durbin–Watson and the Ljung–Box tests reject the hypothesis of uncorrelated residuals. Similarly, the Komlogorov–Smirnov and Jarque–Berra statistics indicate the absence of normality in the residuals, which is logical taking into account the generating process of the residuals. Finally, both the R^2 and the $\bar{R}^2$ show a good fit of the network to the data.

Error Criteria   Various error criteria are presented in Table 6.2. These criteria will help us assess the fit of the wavelet network to the data: in other words, how well the wavelet network learned the data. All criteria are very small, with MSE and NMSE 0.0313 and 0.286, respectively. On the other hand, the MAPE and SMAPE are 119.39% and 44.74%, respectively. Note that there is a large presence of noise to the data. As a result, the error criteria will increase. The wavelet network learned the underlying function successfully as presented in earlier chapters without being affected by the noise.

Scatter Plot   Figure 6.3 is a scatter plot of the real values versus the values forecasted

TABLE 6.2   Case 1: Error Criteria[a]

Md.AE	MAE	MaxAE	SSE	RMSE	NMSE	MSE	MAPE	SMAPE
0.0866	0.1276	0.6813	31.3626	0.1770	0.2896	0.0313	119.39%	44.74%

[a]Md.AE, median absolute error; MAE, mean absolute error; MaxAE, maximum absolute error; SSE, sum of squared errors; RMSE, root mean square error; NMSE, normalized mean squared error; MSE, mean square error; MAPE, mean absolute percentage error; SPAME, symmetric mean absolute percentage error.

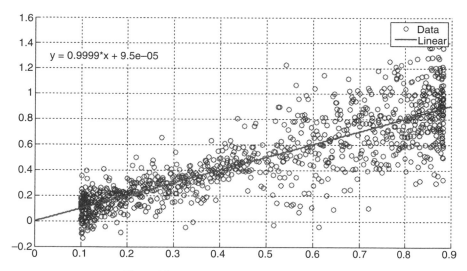

Figure 6.3 *Scatter plot for the first simulated case.*

for the wavelet network. As can be seen, the forecasts and the targets are in line, but there is a large dispersion, caused by the large noise term.

Regression Between the Forecasts and the Target Values The estimated parameters of the regression between the forecasted values and the target values are presented in Table 6.3. Close inspection reveals that parameter b_0 is not statistically different from zero, while the parameter b_1 is not statistically different from 1. Moreover, the linear regression is statistically significant, according to the F-statistic.

Changes in Direction The POCID, IPOCID, and POS criteria are presented in Table 6.4. In this application the aim is to learn the underlying process that is masked by a large noise part with increasing variance. As a result, we expect the POCID and IPOCID to be relatively low. More precisely, the POCID and IPCOID are 75.4% and 50.9%, respectively, while the POS is 97.5%.

TABLE 6.3 **Case 1: Regression Statistics for the First Simulated Case**[a]

	Parameter	p-Values	S.E.	T-Stat.
b_0	0.0001	0.9934	0.0114	0.0083
b_1	1.0000	0.0000	0.0202	49.5080
$b_1 = 1$ test	1.0000	0.9996	0.0202	−0.0004
R^2	0.7104			
F	2451.1000	0.0000		
DW	1.9594	0.49959		

[a]S.E., squared errors; DW, Durbin–Watson.

TABLE 6.4   Case 1: Change-in-Direction Metrics[a]

POCID	IPOCID	POS
75.4%	50.9%	97.5%

[a]POCID, prediction of change in direction; IPOCID, independent prediction of change in direction; POS, prediction of sign.

TABLE 6.5   Case 1: Out-of-Sample Residual Testing[a]

	Parameter	p-Values
n/p Ratio	37.5000	
Mean	0.0031	
Median	−0.0089	
S. dev.	0.2211	
DW	1.8484	0.1876
LB Q-stat.	36.7980	0.0124
JB stat.	5.2543	0.0617
KS stat.	5.3766	0.0000
R^2	0.6578	
$\bar{R}^2$	0.6496	

[a]S. dev., standard deviation; DW, Durbin–Watson; LB, Ljung–Box; KS, Kolmogorov–Smirnov.

Out-of-Sample Forecasts   In this section the out-of-sample performance of the wavelet network is evaluated. The various descriptive statistics as well as various tests on the out-of-sample residuals are presented in Table 6.5. The mean of the residuals is close to zero with a standard deviation of 0.22. Both the Durbin–Watson and Ljung–Box tests reject the hypothesis of uncorrelated residuals. Similarly, the Komlogorov–Smirnov test indicates the absence of normality in the residuals. On the other hand, the normality hypothesis is accepted at a 5% confidence level using the Jarque–Berra statistic. Hence, comparing Tables 6.1 and 6.5, we can conclude that there are similar circumstances that generated the training and the validation sample. Finally, both R^2 and $\bar{R}^2$ show the good prediction ability of the network, considering the large amount of noise.

Next, the accuracy of the prediction is evaluated. In Table 6.6 various error criteria are presented. The values of the metrics are similar to those presented in Table 6.2. The MSE is 0.0487, indicating the very good forecasting ability of the wavelet network. On the other hand, MAPE and SMAPE are relatively high, presenting the high impact of the large noise term.

TABLE 6.6   Case 1: Out-of-Sample Error Criteria[a]

Md.AE	MAE	MaxAE	SSE	RMSE	NMSE	MSE	MAPE	SMAPE
0.1243	0.1669	0.7171	14.6254	0.2208	0.3422	0.0487	145.67%	62.59%

[a]Md.AE, median absolute error; MAE, mean absolute error; MaxAE, maximum absolute error; SSE, sum of squared errors; RMSE, root mean squared error; NMSE, normalized mean squared error; MSE, mean squared error; MAPE, mean absolute percentage error; SPAME, symmetric mean absolute percentage error.

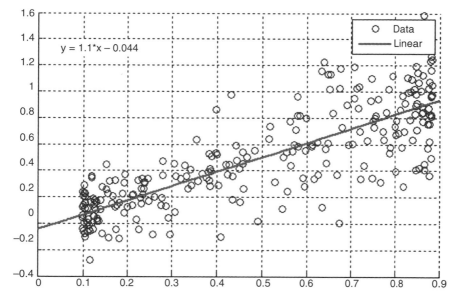

$$y = 1.1^*x - 0.044$$

Figure 6.4   *Case 1: out-of-sample scatter plot.*

The scatter plot of the out-of-sample forecasts and the target values is presented in Figure 6.4, and the estimated parameters of the linear regression between the values forecasted and the target values are presented in Table 6.7. A close inspection of the table reveals that the parameter b_0 is not statistically different from zero, while the parameter b_1 is not statistically different from 1 at a confidence level of 5%. Moreover, the linear regression is statistically significant according to the F-statistic. Finally, as presented in Table 6.8, POCID, IPOCID, and POS are 65.9%, 45.5%, and 92.3%, respectively.

TABLE 6.7   Case 1: Out-of-Sample Regression Statistics[a]

	Parameter	p-Values	S.E.	T-Stat.
b_0	−0.044029	0.082175	0.025245	−1.744089
b_1	1.08430	0.000000	0.044898	24.150415
$b_1 = 1$ test	1.08430	0.061413	0.044898	1.877597
R^2	0.661841			
F	583.243	0.000000		
DW	1.870093	0.235214		

[a]S.E., squared error; DW, Durbin–Watson.

TABLE 6.8   Case 1: Change-in-Direction Metrics[a]

POCID	IPOCID	POS
65.9%	45.5%	92.3%

[a]POCID, prediction of change in direction; IPOCID, independent prediction of change in direction; POS, prediction of sign.

TABLE 6.9   Case 2: Residuals Testing[a]

	Parameter	p-Values
n/p Ratio	38.5000	
Mean	0.0002	
Median	0.0004	
S. dev.	0.0661	
DW	1.9415	0.3540
LB Q-stat.	10.9627	0.9472
JB stat.	908,651.8641	0.0000
KS stat.	14.4117	0.0000
R^2	0.9974	
$\bar{R}^2$	0.9973	

[a]S.dev., standard deviation; DW, Durbin–Watson; LB, Ljung–Box; KS, Kolmogorov–Smirnov.

Case 2: Sum of Sinusoids and Cauchy Noise

As discussed in Chapters 3 and 4, a wavelet network with 8 hidden units was used. This simulated case incorporated large outliers to the data. In this section we examine how well the wavelet network was trained on the data. In other words, we assess the ability of the network to learn the data and then forecast the future values of the underlying function.

Testing the Residuals   First, the residuals of the trained networks are examined. The various descriptive statistics as well as various tests on the residuals are presented in Table 6.9. The mean of the residuals is close to zero, 0.002, while the standard deviation is only 0.0661. The Ljung–Box tests reject the hypothesis of correlation on the residuals. The Komlogorov–Smirnov and the Jarque–Berra statistics indicate the absence of normality in the residuals, which is logical given the fact the residuals were generated by a Cauchy process. Finally, both the R^2 and the $\bar{R}^2$ show a good fit of the network to the data, with values over 99.7%.

Error Criteria   Various error criteria are presented in Table 6.10. These criteria will help us assess the fit of the wavelet network to the data: in other words, how well the wavelet network learned the data. A closer inspection of the table reveals that all criteria are very small, indicating a very good fit of the wavelet network to the data. The MSE and NMSE 0.0044 and 0.0026, respectively. On the other hand, MAPE and SMAPE are only 3.80% and 2.03%, respectively.

TABLE 6.10   Case 2: Error Criteria[a]

Md.AE	MAE	MaxAE	SSE	RMSE	NMSE	MSE	MAPE	SMAPE
0.0084	0.0201	1.1068	4.3683	0.0661	0.0026	0.0044	3.80%	2.03%

[a]Md.AE, median absolute error; MAE, mean absolute error; MaxAE, maximum absolute error; SSE, sum of squared errors; RMSE, root mean squared error; NMSE, normalized mean squared error; MSE, mean squared error; MAPE, mean absolute percentage error; SPAME, symmetric mean absolute percentage error.

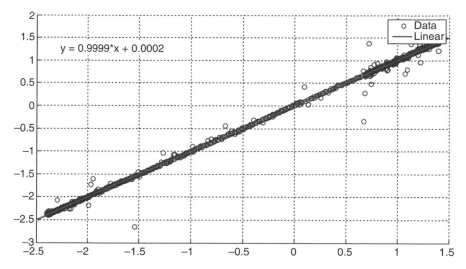

Figure 6.5 *Scatter plot for the second simulated case.*

Scatter Plot   Figure 6.5 is a scatter plot of the real values versus the values forecasted for the wavelet network. As can be seen, the forecasts and the targets are in line.

Regression Between the Forecasts and the Target Values   The estimated parameters of the regression between the forecasted values and the target values are presented in Table 6.11. A close inspection reveals that the parameter b_0 is not statistically different from zero, while the parameter b_1 is not statistically different from 1. Moreover, the linear regression is statistically significant according to the F-statistic.

Changes in Direction   The POCID, IPOCID, and POS criteria are presented in Table 6.12. In this application the aim for the wavelet network is to learn the underlying process in the presence of large outliers. The POCID, IPOCID, and POS are 81.40%, 75.10%, and 99.80%, indicating that the wavelet network can successfully identify and predict the changes in the direction of the underlying function.

TABLE 6.11   Case 2: Regression Statistics for the Second Simulated Case[a]

	Parameter	p-Values	S.E.	T-Stat.
b_0	0.0002	0.9147	0.0021	0.1072
b_1	0.9999	0.0000	0.0016	619.8138
$b_1 = 1$ test	0.9999	0.9577	0.0016	−0.0531
R^2	0.9974			
F	384,170.0000	0.0000		
DW	1.9415	0.338		

[a]S.E., squared error; DW, Durbin–Watson.

TABLE 6.12   Case 2: Change-in-Direction Metrics[a]

POCID	IPOCID	POS
81.40%	75.10%	99.80%

[a]POCID, prediction of change in direction; IPOCID, independent prediction of change in direction; POS, prediction of sign.

TABLE 6.13   Case 2: Out-of-Sample Residuals Testing[a]

	Parameter	p-Values
n/p Ratio	11.5385	
Mean	0.0043	
Median	−0.0008	
S. dev.	0.0655	
DW	2.0068	0.9523
LB Q-stat.	15.3770	0.7544
JB stat.	143,397.8894	0.0000
KS stat.	7.9658	0.0000
R^2	0.9971	
$\bar{R}^2$	0.9969	

[a]S. dev., standard deviation; DW, Durbin–Watson; LB, Ljung–Box; KS, Kolmogorov–Smirnov.

Out-of-Sample Forecasts   In this section the out-of-sample performance of the wavelet network is evaluated. The various descriptive statistics as well as various tests on the out-of-sample residuals are presented in Table 6.13. The mean of the residuals is close to zero with a standard deviation of 0.06. The Ljung–Box test accepts the hypothesis of uncorrelated residuals. Similarly, the Komlogorov–Smirnov and Jarque–Berra tests indicate the absence of normality in the residuals. Finally, both R^2 and $\bar{R}^2$ are over 99.6%, showing the good predictive ability of the network.

Next, the accuracy of the prediction is evaluated. Various error criteria are presented in Table 6.14. The values of the metrics are similar to those presented in Table 6.10. MSE, NMSE, and RMSE are 0.0043, 0.0029, and 0.0656, indicating the very good forecasting ability of the wavelet network. On the other hand, MAPE and SMAPE are only 3.88% and 0.19%.

Figure 6.6 is a scatter plot of the out-of-sample forecasts and the target values. The estimated parameters of the linear regression between the values forecasted and the target values are presented in Table 6.15. A close inspection of the table

TABLE 6.14   Case 2: Out-of-Sample Error Criteria[a]

Md.AE	MAE	MaxAE	SSE	RMSE	NMSE	MSE	MAPE	SMAPE
0.0074	0.0198	0.8572	1.2903	0.0656	0.0029	0.0043	3.88%	0.19%

[a]Md.AE, median absolute error; MAE, mean absolute error; MaxAE, maximum absolute error; SSE, sum of squared errors; RMSE, root mean squared error; NMSE, normalized mean squared error; MSE, mean squared error; MAPE, mean absolute percentage error; SPAME, symmetric mean absolute percentage error.

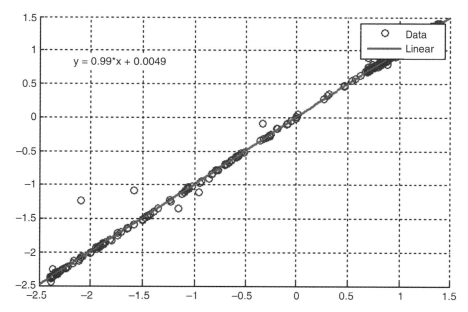

Figure 6.6 *Case 2: out-of-sample scatter plot.*

TABLE 6.15   Case 2: Out-of-Sample Regression Statistics[a]

	Parameter	p-Values	S.E.	T-Stat.
b_0	0.0049	0.1928	0.0038	1.3052
b_1	0.9944	0.0000	0.0031	324.9527
$b_1 = 1$ test	0.9944	0.0664	0.0031	−1.8425
R^2	0.9972			
F	105,594.2273	0.0000		
DW	2.0384	0.7830		

[a]S.E., squared error; DW, Durbin–Watson.

reveals that the parameter b_0 is not statistically different from zero and the parameter b_1 is not statistically different from 1 at a confidence level of 5%. Moreover, the linear regression is statistically significant according to the F-statistic. Finally, as presented in Table 6.16, POCID, IPOCID, and POS are 82.29%, 79.93%, and 99.33%, respectively.

TABLE 6.16   Case 2: Change-in-Direction Metrics[a]

POCID	IPOCID	POS
86.29%	79.93%	99.33%

[a]POCID, prediction of change in direction; IPOCID, independent prediction of change in direction; POS, prediction of sign.

CLASSIFICATION

In classification applications, the primary objective is to determine the group to which an "object" (person, company, product, etc.) belongs (Green and Carroll, 1978). In finance, some of the usual classification applications are: prediction of the success or failure of a new product, determination of credit risk of client, and bankruptcy probability.

In statistical terminology such applications are known as *applications of discriminant analysis*. Discriminant analysis is the appropriate statistical approach when the dependent variable is categorical—nominal or nonmetric—while the independent variables are numerical.

If the dependent variable is composed of two groups or classifications, it is called *discriminant analysis of two groups*. When the dependent variable is composed of three or more groups or classifications, it is called *multiple discriminant analysis*.

Discriminant analysis involves extraction of the linear combination of dependent variables that will better distinguish among the a priori fixed groups. The linear combinations for a discriminant analysis are derived from an equation that takes the following form:

$$z = w_1 x_1 + w_2 x_2 + \cdots + w_m x_m \tag{6.47}$$

where $i = 1, \ldots, m$, x_i are the independent variables, w_i are the discriminant weights, and z is the discriminant score. In the case of wavelet networks, the linear relationship (6.47) is replaced by the following nonparametric relationship:

$$z = g_\lambda \left(\mathbf{x}; \hat{\mathbf{w}}_n \right) \tag{6.48}$$

From equation (6.47) it can be derived that in discrete analysis each independent variable is multiplied by its respective weight and then the products are added to estimate the discrete score of vector x_i. The average value of the discrete scores of all individuals within a given group is called a *centroid*. The number of centroids is equal to the number of groups. For simplicity, we restrict our analysis to applications with only two centroids. The greater the distance between the centroids, the smaller the overlap of the distributions of distinct scores of the two groups and therefore the better the distinct function.

A distinct function that provides good separation between classes A and B is presented in Figure 6.7. On the other hand, a distinct function that performs a relatively poor separation between classes A and B is presented in Figure 6.8.

Assumptions and Objectives of Discriminant Analysis

The basic assumptions under the (linear) discriminant analysis is a multivariate normality of the independent variables and unknown (but equal) dispersion and covariance matrices for the groups as defined by the dependent variable (Green and Carroll, 1978; Harris, 2001).

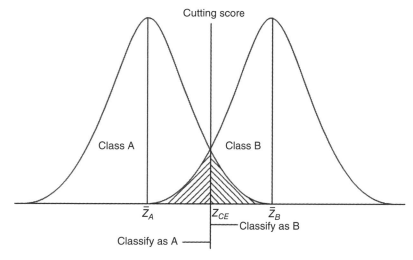

Figure 6.7 *Good separation between groups A and B and the optimal cutoff value for samples of the same size.*

Additionally, it must be ensured that the independent variables are truly linearly independent and that multicollinearity between the input variables does not exist. This assumption becomes especially important when stepwise variable selection methods, such as the ones presented in Chapter 5, are used (Hair et al., 2010).

Finally, regarding the size of the sample (either of the training or of the validation sample), a general rule that many studies suggest is that the sample must contain at least 20 observation for each independent variable (Hair et al., 2010).

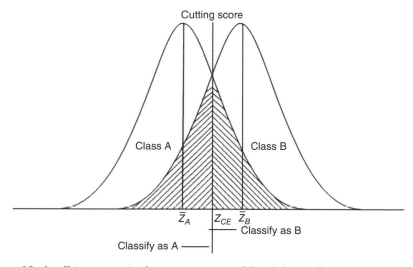

Figure 6.8 *Insufficient separation between groups A and B and the cutoff value for samples of the same size.*

The objectives of discriminant analysis are summarized as follows:

- To create models for the classification of individuals or objects into groups on the basis of their scores on several variables.
- To determine whether statistically significant differences exist between the average score profiles of the two (or more) a priori defined groups.
- To determine the independent variables that contribute most in the average discriminant score.

It is clear from the foregoing objectives that discriminant analysis is useful in the correct classification of statistical units into groups or classes or in an understanding of the differences among the various groups. Hence, discriminant analysis can be used either as a predictive technique or as a type of profile analysis (Hair et al., 2010). In this book we are interested in using discriminant analysis as a predictive technique.

Validation of the Discriminant Function

The next step in discriminant analysis is the validation. After estimation of the discriminant function, the statistical significance of the function must be estimated. This can be done either by the calculation of chi-square or the Mahalanobis D^2 statistics. Although these statistics assess the significance of the discriminant function, they do not provide any information about the predictive power of the function.

On the other hand, the predictive power of the discriminant function can be determined by the classification matrix. The classification matrix can be related to the concept of the R^2 in linear regression. As in the case of linear regression, there are cases where the regression is statistically significant; however, the R^2 is very low. In other words, the linear regression explains a very small percentage of the variance. Similarly, the discriminant function can be statistically significant but with low predictive ability. In discriminant analysis the percentage of the cases classified correctly is called the *hit ratio* and is analogous to R^2.

Cutting Score Determination   Before proceeding with construction of the classification matrix, we need to determine the cutting score. The *cutting score criterion* is the score against which each case's discriminant score is compared to determine into which group the observation should be classified. To construct the classification matrix, the optimal cutting score or critical Z values must be determined. The optimal cutting score will differ depending on whether the sizes of the groups are equal or unequal.

If the two groups are of equal size, that is, have the same number of observations, the optimal cutting score will be located in the middle of the distance between the two centroids. In this case the cutting score is given by

$$z = \frac{\bar{z}_A + \bar{z}_B}{2} \tag{6.49}$$

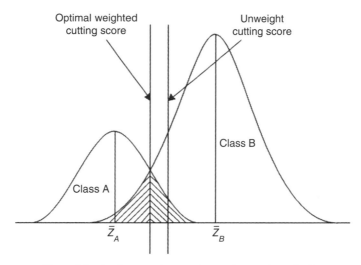

Figure 6.9 *Cutting score for two classes of unequal sample size.*

where z is the critical cutting score value for equal group sizes, $\bar{z}_A$ is the centroid for group A, and $\bar{z}_B$ is the centroid for group B.

If the two groups are not of equal size, the cutting score is estimated by a weighted average as follows:

$$z = \frac{n_A \bar{z}_A + n_B \bar{z}_B}{n_A + n_B} \tag{6.50}$$

where z is the critical cutting score value for unequal group sizes, $\bar{z}_A$ is the centroid for group A, $\bar{z}_B$ is the centroid for group B, n_A is the number of observations in group A, and n_B is the number of observations in group B. An example of a cutting score for two groups of unequal size is presented in Figure 6.9. Note that formula (6.49) is a special case of (6.50) with $n_A = n_B$. Furthermore, in formulas (6.49) and (6.50), a normal distribution and a known covariance are assumed.

In determining the optimal cutoff score, the cost of classifying a case incorrectly should be taken into account. If the costs of an incorrect prediction in various cases are about the same, the optimal cutoff score will be the one that will classify incorrectly the smaller number of cases in all groups. If the costs of misclassifications are not the same, the optimal cutoff score will be the one that minimizes the total cost of the misclassification. For an in-depth analysis in defining the optimum cutoff score, we refer to Dillon and Goldstein (1984) and Huberty et al. (1987).

For construction of the classification matrix, for each case in which it is included in the testing sample, its discriminant score must be compared against the cutting score. Then each case is classified as follows:

- Classify case n intro group A if $z_n < z_{cs}$.
- Classify case n intro group B if $z_n > z_{cs}$.

TABLE 6.17   Classification Matrix

Target	Forecast 0	Forecast 1	Total
0	α	β	$\alpha + \beta$
1	γ	δ	$\gamma + \delta$
Total	$\alpha + \gamma$	$\beta + \delta$	$\alpha + \beta + \gamma + \delta$

Here z_n is the discriminant score of observation n and z_{cs} is the critical cutting score value.

The results of the classification are represented in matrix form. Table 6.17 shows an example of a classification matrix.

The significance of the classification accuracy can be determined by the following t-test in the case of two equal sample groups:

$$t = \frac{p - 0.5}{\sqrt{0.5(1 - 0.5)/n}} \tag{6.51}$$

where p is the proportion classified correctly and n is the sample size. Formula (6.51) can easily be modified to be used with more groups and unequal sample sizes.

Evaluating the Classification Ability of a Wavelet Network

In the preceding section, the classification accuracy of a discriminant function was measured by the hit ratio, which is obtained by the classification matrix. Additional measures that can describe the predictive and classification accuracy are presented in this section.

Relative Entropy   The first measure presented, the relative entropy, can be estimated as follows:

$$RE = -\frac{1}{n}\left[\sum_{s=1}^{n_A} \ln\left(1 - y_A^{(s)}\right) + \sum_{t=1}^{n_B} \ln\left(y_B^{(t)}\right)\right] \tag{6.52}$$

where n_A and n_B are the observations in class A (0) and B (1), respectively, and $y_A^{(s)}$ and $y_B^{(t)}$ are the outputs of the network for the patterns s in class 0 and t in class 1. If the relative entropy is close to zero, it indicates a good fit to the data.

Kolmogorov–Smirnov Statistic   Another indicator of the classification power of a wavelet network that can be used is the Kolmogorov–Smirnov (KS) statistic. To estimate the KS statistic, cumulative histograms of the projected outputs of the network are computed and their overlaps are estimated.

The KS statistic is the maximum distance between the respective percentile values of the two histograms. If the output values of the model were random, the cumulative histograms would both be linear and the KS statistic would be close to zero. If the model were perfect, the KS statistic would be equal to 1.

Maximum Chance Criterion   Before we proceed with our analysis, we must define what is considered an acceptable level of prediction for a distinct function. This depends on the percentage number of observations that can be classified correctly in a random manner. If every class consists of the same number of observations, the percentage is 1 divided by the number of classes.

However, when the number of observations is different for each class, the maximum chance criterion is used. This criterion is determined by the largest class. For example, if the data set consists of two classes, where the first has 65 observations and the second has 35, the maximum chance criterion is 65%. In other words, if our model cannot classify correctly at least 65% of the observations, it does not have any classification power.

Proportional Chance Criterion   The preceding criterion should be used when the sole objective aim of the analysis is to maximize the proportion of cases classified correctly. However, we are usually interested in the correct classification in both classes. In other words, the criterion above does not take into account the classification in the smaller group. Returning to the preceding example, where class A (0) had 35 observations and class B (1) had 65 observation. If we classify all cases (all of the 100 cases) as 1, our strategy will produce 65 correct classifications.

Alternatively, the proportional chance criterion can be used:

$$PRO = p^2 + (1 - p)^2 \qquad (6.53)$$

where p is the percentage of the cases that belong to class A and $1 - p$ is the percentage of the cases that belong to class B.

Classification Accuracy Relative to Chance   A crucial question that must be asked is whether our model has a percentage of correct classification significantly larger than would be expected by chance. If the wavelet network has classification accuracy greater than the one that can be expected by chance, we can proceed on the interpretation of the discriminant functions. Otherwise, an analysis would not provide meaningful results (Hair et al., 2010).

The following quick criterion is presented by Hair et al. (2010): The classification accuracy should be at least 25% greater than that achieved by chance. For example, if chance accuracy is 50%, the classification accuracy should be at least 62.5%.

A more robust test is Press's Q-statistic. This measure compares the number of correct classifications with the total sample size and the number of groups. The value calculated is then compared with a critical value obtained for chi-square with 1 degree of freedom. If the estimated value is greater than the critical value, a significantly better classification ability can be obtained using the model than by chance.

Press's Q-statistic is given by

$$Q = \frac{[n - (c \times k)]^2}{n(k-1)} \tag{6.54}$$

where n is the sample size, c the number of correct classifications, and k the number of classification groups. This test is sensitive to sample size, where large samples are more likely to show significance than small sample sizes of the same classification rate.

Sensitivity   The sensitivity test measures the ability of a model to identify positive results. The sensitivity, also called the *true positive rate*, is given by the relationship

$$\text{sensitivity} = \frac{\text{true positives}}{\text{true positives} + \text{false negatives}} = \frac{\text{true positives}}{\text{total positives}} \tag{6.55}$$

Specificity   Similarly, the specificity test measures the ability of a model to identify negative results. The specificity, also called the *true negative rate*, is given by the following relationship:

$$\text{specificity} = \frac{\text{true negatives}}{\text{true negatives} + \text{false positives}} = \frac{\text{true negatives}}{\text{total negatives}} \tag{6.56}$$

Rate of Correctness (Hit Ratio)   As already presented, the rate of correctness or hit ratio measures the percentage of the correct classification of the wavelet network:

$$\text{RC} = \frac{\text{true positives} + \text{true negatives}}{\text{true positives} + \text{true negatives} + \text{false positives} + \text{false negatives}} \tag{6.57}$$

Rate of Missing Chances   The rate of missing chance measures the percentage of true cases that were classified as false:

$$\text{RMC} = \frac{\text{false negatives}}{\text{false negatives} + \text{true positives}} \tag{6.58}$$

Rate of Failure   The rate of failure measures the percentage of false cases identified as true.

$$\text{RF} = \frac{\text{false positives}}{\text{false positives} + \text{true negatives}} \tag{6.59}$$

Fitness Functions   The criteria above can be used to construct fitness functions. Fitness functions are useful for comparing different models. An example of a fitness function is

$$\text{fitness} = 0.6\,\text{RC} - 0.1\,\text{RMC} - 0.3\,\text{RF} \tag{6.60}$$

In equation (6.60), different weights can be used, depending on the application.

Case 3: Classification Example on Bankruptcy

In this section a different problem is considered. A wavelet network is constructed to classify if a firm will or will not go bankrupt based on various attributes. The data set contains samples from 240 Greek firms. Each instance has one of two possible classes: bankrupt or nonbankrupt. The two groups have the same number of firms; hence, there are 120 firms in each group. The objective is to construct a wavelet network that accurately classifies each firm. The classification is based on four attributes proposed by Altman and Saunders (1997):

$$X_1 = \frac{\text{working capital}}{\text{total assets}}$$

$$X_2 = \frac{\text{retained earnings}}{\text{total assets}}$$

$$X_3 = \frac{\text{earnings before interest and tax}}{\text{total assets}}$$

$$X_4 = \frac{\text{book value of equity}}{\text{total liabilities}}$$

Our results will be compared against a linear classification model.

The data were split randomly into training and validation samples. The training sample consists of 168 (70%) cases. The validation sample consists of 72 (30%) cases and is used to evaluate the predictive and classification power of the trained wavelet network.

In the training sample, 86 firms went bankrupt and were given the value 0, and 82 firms that were not bankrupt and were given the value 1. The cutting score is 0.4882. A classification matrix of the training sample using a wavelet network is presented in Table 6.18. A closer inspection of the table reveals very good classification rates. More specifically, the wavelet network classified correctly 56 nonbankrupt firms and 70 bankrupt cases. Hence, the wavelet network classified correctly 126 of 168 cases (75%). The specificity of the models is 68.29% and the sensitivity is 81.40%. On the other hand, the rate of failure and the rate of missing chances are 18.60% and 31.71%, respectively. Finally, the fitness function given by (6.60) is 0.3625.

Similarly, in Table 6.19 the classification matrix is presented when a linear model is used. A closer inspection reveals that the classification ability of the linear model in-sample is worse. The hit ratio is only 70.83% and the sensitivity and specificity

TABLE 6.18 In-Sample Classification Matrix: Wavelet Network

	Forecast				
Target	Nonbankrupt	Bankrupt	Total	Sensitivity	Specificity
Nonbankrupt	56	26	82	68.29%	81.40%
Bankrupt	16	70	86	Rate of Missing Chances	Rate of Failure
Total	72	96	168	31.71%	18.60%

TABLE 6.19   In-Sample Classification Matrix: Linear Case

Target	Forecast Bankrupt	Forecast Nonbankrupt	Total	Sensitivity	Specificity
Bankrupt	55	27	82	67.07%	74.42%
Nonbankrupt	22	64	86	Rate of Missing Chances	Rate of Failure
Total	77	91	168	32.93	25.58%

TABLE 6.20   Evaluation of the Classification Ability of the Wavelet Network

Maximum Chance	1.25% Max. Chance	Pro	Press's Q^a	Hit Ratio
51.19%	63.99%	50.03%	42	75.00%

[a]Press's Q critical values at confidence levels 0.1, 0.05, and 0.01: 2.71, 3.84, 6.63.

are 67.07% and 74.42%, respectively. Finally, the fitness function is reduced and is only 0.3153.

The evaluation of the classification ability of the wavelet network is presented in Table 6.20. The maximum chance criterion is 51.19% and the heuristic method presented by Hair et al. (2010) is 63.99%. Finally, the proportional chance criterion is 51.77%. The hit ratio is significantly larger than the various chance criteria and is 75%. Hence, the model is predicting significantly better than chance. This is also confirmed by the large value of Press's Q-statistic, which is greater than the critical values at confidence levels 0.1, 0.05, and 0.01. Similarly, the evaluation of the classification ability of the linear model is presented in Table 6.21. The hit ratio is 70.83% and Press's Q-statistic is higher than the critical values. Hence, both the wavelet network and the linear model predict significantly better than chance.

Out-of-Sample   Next, the forecasting and classification ability of the trained wavelet network is evaluated in the validation sample. The data of the validation sample were not used during the training phase. Hence, these are new data that were never presented to the wavelet network.

In the validation sample there are 40 bankrupt cases given the value 0 and 32 nonbankrupt cases given the value 1. The classification matrix of the training sample using the wavelet network is presented in Table 6.22. A closer inspection of the table reveals very good prediction ability and classification rates. More specifically, the wavelet network classified correctly 24 nonbankrupt cases and 31 bankrupt cases. Hence, the wavelet network classified correctly 55 of 72 cases (76.39%). The specificity of the models is 77.50% and the sensitivity is 75%. On the other hand, the rate

TABLE 6.21   Evaluation of the Classification Ability of the Linear Model

Maximum Chance	1.25% Max. Chance	Pro	Press's Q^a	Hit Ratio
51.19%	63.99%	50.03%	29.16	70.83%

[a]Press's Q critical values at confidence levels 0.1, 0.05, and 0.01: 2.71, 3.84, 6.63.

TABLE 6.22   Out-of-Sample Classification Matrix: Wavelet Network

| Target | Forecast | | | Sensitivity | Specificity |
	Nonbankrupt	Bankrupt	Total		
Nonbankrupt	24	8	32	75.00%	77.50%
Bankrupt	9	31	40	Rate of Missing Chances	Rate of Failure
Total	33	39	72	25.00%	22.50%

TABLE 6.23   Out-of-Sample Classification Matrix: Linear Case

| Target | Forecast | | | Sensitivity | Specificity |
	Nonbankrupt	Bankrupt	Total		
Nonbankrupt	23	9	32	71.88%	65.00%
Bankrupt	14	26	40	Rate of Missing Chances	Rate of Failure
Total	37	35	72	28.12%	35.00%

of failure and the rate of missing chances are low, 22.50% and 25%, respectively. Finally, the fitness function is 0.3658.

The out-of-sample classification matrix of the linear model is presented in Table 6.23. It is clear that the classification accuracy is reduced when a linear model is used. More precisely, the sensitivity and specificity were reduced to 71.88% and 65% while the linear model classified correctly only 49 of the 72 cases. Similarly, the rate of missing chances and the rate of failure were increased to 28.12% and 35%. Finally, the fitness function is 0.2752.

The evaluation of the classification ability of the wavelet network and the linear model are presented in Tables 6.24 and 6.25, respectively. The maximum chance criterion is 55.56% and the heuristic method presented by Hair et al. (2010) is 69.44%. Finally, the proportional chance criterion is 50.62%. The hit ratio of the wavelet network is significantly larger than the maximum chance and the proportional chance criteria and is 76.39%. Similarly, the hit ratio of the linear model is 68.06%. Hence, both the linear and the wavelet network models are predicting significantly better than chance. This is confirmed by the large value of Press's Q-statistic, which

TABLE 6.24   Out-of-Sample Evaluation of the Classification Ability of the Wavelet Network

Maximum Chance	1.25% Max. Chance	Pro	Press's Q^a	Hit Ratio
55.56%	69.44%	50.62%	20.05	76.39%

[a]Press's Q critical values at confidence levels 0.1, 0.05, and 0.01: 2.71, 3.84, 6.63.

TABLE 6.25   Out-of-Sample Evaluation of the Classification Ability of the Linear Model

Maximum Chance	1.25% Max. Chance	Pro	Press's Q^a	Hit Ratio
55.56%	69.44%	50.62%	9.39	68.06%

[a]Press's Q critical values at condidence levels 0.1, 0.05, and 0.01: 2.71, 3.84, 6.63.

is greater than the critical values at confidence levels 0.1, 0.05, and 0.01. However, it is clear that the wavelet network outperforms the linear model. Not only is the hit ratio significantly higher when a wavelet network is used, but the rate of failure, which leads to false decisions and loss of money, is also significantly smaller.

CONCLUSIONS

In this chapter various metrics for measuring model adequacy and the predicting ability of a wavelet network were presented. First, various tests were presented that test the properties of the residuals of the fitted wavelet network. Next, the forecasting ability of a wavelet network was tested using scatter plots and a linear regression between the values forecasted obtained from the wavelet network and the target values. Depending on the application, we may be interested in three categories of predictions. In the first case, we are interested in value forecasting. In this case, the predictive power of the wavelet network is measured as an expression of the difference between the target values and the output of the network. Second, in classification applications we are interested in the correct classification of various cases. In this case the predictive power of the network is measured as the ability of the wavelet network to classify the individual cases correctly. Finally, there are applications where we are interested only in the sign or in the change in the direction of the values. In this case, metrics such as the POS and IPOCID are used.

In this section the wavelet network models were tested for adequacy in three cases. The first two were time-series approximation and forecasting problems; the last was a classification problem. More precisely, in the last case study, a wavelet network was constructed and used to classify the credit risk of firms. Our results indicate that the nonlinear nonparametric wavelet network outperforms the linear model significantly.

REFERENCES

Altman, E. I., and Saunders, A. (1997). "Credit risk measurement: developments over the last 20 years." *Journal of Banking and Finance*, 21(11–12), 1721–1742.

Box, G. E. P., and Jenkins, G. M. (1970). *Time Series Analysis, Forecasting and Control*, Holden-Day, San Francisco.

Dillon, W. R., and Goldstein, M. (1984). *Multivariate Analysis*. Wiley, New York.

Durbin, J., and Watson, G. S. (1951). "Testing for serial correlation in least squares regression: II." *Biometrika*, 38(1/2), 159–177.

Green, P. E., and Carroll, J. D. (1978). *Analyzing Multivariate Data*. Dryden Press, Hinsdale, IL.

Hair, J., Black, B., Babin, B., and Anderson, R. (2010). *Multivariate Data Analysis*, 7th ed. Prentice Hall, Upper Saddle River, NJ.

Harris, R. J. (2001). *A Primer of Multivariate Statistics*. Lawrence Erlbaum Associates, Hillsdale, NJ.

Huberty, C. J., Wisenbaker, J. M., and Smith, J. C. (1987). "Assessing predictive accuracy in discriminant analysis." *Multivariate Behavioral Research*, 22(3), 307–329.

Judge, G., Hill, R. C., Griffiths, W. E., Lutkepohl, H., and Lee, T.-C. (1982). *Introduction to the Theory and Practice of Econometrics*. Wiley, New York, p. 880.

Makridakis, S., Wheelwright, S. C., and Hyndman, R. J. (2008). *Forecasting Methods and Applications*. Wiley, Hoboken, NJ.

Ramsey, J. B. (1969). "Tests for specification errors in classical linear least-squares regression analysis." *Journal of the Royal Statistical Society*, Seri B, 350–371.

White, H. (1989). "Learning in artificial neural networks: a statistical perspective." *Neural Computation*, 1, 425–464.

Yao, J., Tan, C. L., and Poh, H.-L. (1999). "Neural networks for technical analysis: a study on KLCI." *International Journal of Theoretical and Applied Finance*, 2(02), 221–241.

Zapranis, A., and Refenes, A. P. (1999). *Principles of Neural Model Indentification, Selection and Adequacy: With Applications to Financial Econometrics*, Springer-Verlag, New York.

7

Modeling Uncertainty: From Point Estimates to Prediction Intervals

In earlier chapters a framework was presented where a wavelet network can efficiently be constructed, initialized, and trained. In this chapter we discuss the reliability of estimates of wavelet networks, since forecasts are characterized by uncertainty due to (1) inaccuracy in the measurements of the training data, and (2) limitations of the model. More precisely, in this chapter the framework proposed is expanded by presenting two methods for estimating confidence and prediction intervals.

The output of the wavelet network is the approximation of the underlying function $f(\mathbf{x})$ obtained from the noisy data. In many applications, and especially in finance, risk managers may be more interested in predicting intervals for future movements of the underlying function $f(\mathbf{x})$ than simply point estimates. For example, financial analysts who want to forecast the future movements of a stock are interested not only in the prices predicted but also in the confidence and prediction intervals. For example, if the price of a stock moves outside the prediction intervals, a financial analyst will take a position in the stock. If the price of the stock is below the lower bound, the stock is traded lower that it should be and a long position must be taken. On the other hand, if the price stock is above the upper bound of the prediction interval, the stock is too expensive and a short position should be taken.

In real data sets the training patterns are usually inaccurate, since they contain noise or they are incomplete due to missing observations. Financial time series especially are dominated by these characteristics. As a result, the validity of the predictions of our model (as well as of any other model) is questioned. The uncertainty that results

Wavelet Neural Networks: With Applications in Financial Engineering, Chaos, and Classification,
First Edition. Antonios K. Alexandridis and Achilleas D. Zapranis.
© 2014 John Wiley & Sons, Inc. Published 2014 by John Wiley & Sons, Inc.

from the data, called the *data noise variance* σ_ε^2, contributes to the total variance of the prediction (Breiman, 1996; Carney et al., 1999; Heskes, 1997; Papadopoulos et al., 2000).

On the other hand, presenting to a trained network new data that were not introduced to the wavelet networks during the training phase, additional uncertainty is introduced to the predictions. Since the training set consists of a finite number of training pairs, the solution $\hat{w}_n$ is likely not to be valid in regions not represented in the training sample (Papadopoulos et al., 2000). In addition, the iterative algorithm that is applied to train a wavelet network may converge to a local minimum of the loss function. This source of uncertainty, which arises from misspecifications in the model or in the parameter selection as well as from limitations of the training algorithm also contributes to the total variance of the prediction, is called the *model variance* σ_m^2 (Papadopoulos et al., 2000).

The model variance and the data noise variance are assumed to be independent. The total variance of the prediction is given by the sum of two variances:

$$\sigma_p^2 = \sigma_m^2 + \sigma_\varepsilon^2 \tag{7.1}$$

To apply wavelet networks in financial applications, a statistical measure of the confidence of the predictions must be derived.

If the total variance of a prediction can be estimated, it is possible to construct confidence and prediction intervals. In the first case we are interested in the difference between the network's output and the true underlying generating process; in the second case we are interested in the difference between the network's output and the value observed. We explore this in the remainder of the chapter.

THE USUAL PRACTICE

In the framework of classical sigmoid neural networks, the methods proposed for constructing confidence and prediction intervals fall into three major categories: the analytical, the Bayesian, and the ensemble network methods. Analytical methods provide good prediction intervals only if the training set is very large (De Veaux et al., 1998). They are based on the assumptions that the noise in the data is independent and identically distributed with mean zero and constant standard deviation. In real problems that hypothesis usually does not hold. As a result, there will be intervals where the analytical method either over- or underestimates the total variance. Finally, with analytical methods the effective number of parameters must be identified, although pruning schemes such as the Irrelevant Connection Elimination scheme can be used to solve this problem. On the other hand, Bayesian methods are computationally expensive techniques that need to be tested further (Zapranis and Refenes, 1999; Ζαπράνης, 1999). Results from Papadopoulos et al. (2000) indicate that the use of Bayesian methods and the increase in the computational burden are not justified by their performance. Finally, analytical and Bayesian methods are computationally complex since the inverse of the Hessian matrix must be estimated, which under certain circumstances can be very unstable.

Finally, ensemble network methods create different versions of the initial network and then combine the outputs to provide constancy to the predictor by stabilizing the high variance of a wavelet network. In ensemble network methods the new versions of the network are generally created using bootstrapping. The only assumption needed is that the wavelet network provides an unbiased estimation of the true regression. Moreover, ensemble networks can handle nonconstant variance. Assuming a constant variance is a simplification of reality. In real data sets the variance changes over time as new data arrive. Similarly, in finance, the variance of daily data changes as new information arrives at the traders. Hence, we suppose that the total variance of the prediction is not constant and is given by

$$\sigma_p^2(\mathbf{x}) = \sigma_m^2(\mathbf{x}) + \sigma_\varepsilon^2(\mathbf{x}) \tag{7.2}$$

The methods most often cited are bagging (Breiman, 1996) and balancing (Carney et al., 1999; Heskes, 1997). In the following sections we adapt these two methods to construct confidence and prediction intervals under the framework of wavelet networks. A framework similar to the one presented by Carney et al. (1999) to estimate the total prediction variance σ_p^2 and to construct confidence and prediction intervals is adapted.

CONFIDENCE AND PREDICTION INTERVALS

Suppose that our set of observations is given by $D_n = (\mathbf{x}_i, y_i)$, $i = 1,\ldots,n$, which verifies the following nonlinear nonparametric wavelet network:

$$y_i = g_\lambda(\mathbf{x}_i; \mathbf{w}_0) + \varepsilon_i. \tag{7.3}$$

where y_i is the output of the wavelet network $g_\lambda(\mathbf{x}_i; \mathbf{w}_0)$ and $\mathbf{w}_0$ represents the true vector of parameters for the specific unknown function $\varphi(\mathbf{x}_i)$, which is estimated by the network. This means that

$$g_\lambda(\mathbf{x}_i; \mathbf{w}_0) \approx \varphi(\mathbf{x}_i) \equiv \mathrm{E}[y_i \mid \mathbf{x}_i] \tag{7.4}$$

Initially, we assume that the error ε_i is distributed independently and identically with zero mean and variance σ_ε^2.

The estimation of the vector $\mathbf{w}_0$ using least squares is given by the vector $\hat{\mathbf{w}}_n$. The vector $\hat{\mathbf{w}}_n$ is estimated by minimizing the sum of squares of the error:

$$SSE = \sum_{i=1}^{n} [y_i - g_\lambda(\mathbf{x}_i; \mathbf{w})]^2 \tag{7.5}$$

For the input vector $\mathbf{x}_i$ and the weight vector of the network $\hat{\mathbf{w}}_n$, the output of the network is

$$\hat{y}_i = g_\lambda(\mathbf{x}_i; \hat{\mathbf{w}}_n) \tag{7.6}$$

Within this context, the concept of "confidence" has a dual meaning. In the first case we are interested in the accuracy of the estimation of the actual but unknown function $\varphi(\mathbf{x}_i)$. Namely, the distribution of the quantity is

$$\varphi(\mathbf{x}_i) - g_\lambda(\mathbf{x}_i; \hat{\mathbf{w}}_n) \equiv \varphi(\mathbf{x}_i) - \hat{y}_i \tag{7.7}$$

referred to as the *confidence interval*. In the second case we are interested in the accuracy of the estimation regarding the network output observed. That relates to the distribution of the quantity

$$y_i - g_\lambda(\mathbf{x}_i; \hat{\mathbf{w}}_n) \equiv y_i - \hat{y}_i \tag{7.8}$$

and is referred to as the *prediction interval*.

Figures 7.1 and 7.2 present the relationship between the differences (7.7) and (7.8). In Figure 7.1 the prediction obtained by the wavelet network has a greater value than the value observed, while in Figure 7.2 the opposite is true. We observe that

$$\begin{aligned} y_i - g_\lambda(\mathbf{x}_i; \hat{\mathbf{w}}_n) &= \{\varphi(\mathbf{x}_i) - g_\lambda(\mathbf{x}_i; \hat{\mathbf{w}}_n)\} + \{y_i - \varphi(\mathbf{x}_i)\} \\ \Leftrightarrow y_i - g_\lambda(\mathbf{x}_i; \hat{\mathbf{w}}_n) &= \{\varphi(\mathbf{x}_i) - g_\lambda(\mathbf{x}_i; \hat{\mathbf{w}}_n)\} + \varepsilon_i \end{aligned} \tag{7.9}$$

or equivalently from relations (7.7) and (7.8),

$$y_i - \hat{y}_i = (\varphi(\mathbf{x}_i) - \hat{y}_i) + \varepsilon_i \tag{7.10}$$

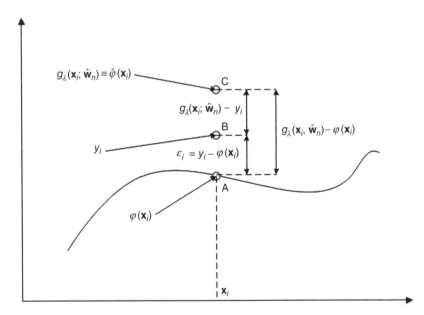

Figure 7.1   *Relationship between the wavelet network output $\hat{y}_i$, the observation y_i, and the underlying function $\varphi(\mathbf{x}_i)$ that created the observation by adding the stochastic term ε_i in the case where the predicted value is greater than the observation ($\hat{y}_i > y_i$).*

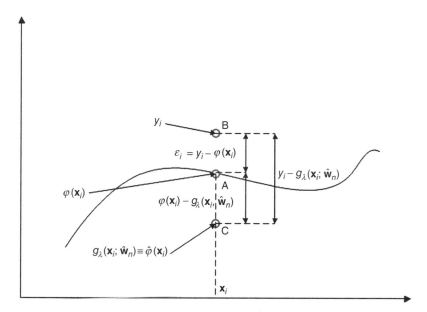

Figure 7.2 *Relationship between the wavelet network output $\hat{y}_i$, the observation y_i, and the underlying function $\varphi(\mathbf{x}_i)$ that created the observation by adding the stochastic term ε_i in the case where the predicted value is less than the observation ($\hat{y}_i > y_i$).*

From (7.10) we can conclude that the confidence interval is included in the prediction interval. Since the difference $(\varphi(\mathbf{x}_i) - \hat{y}_i)$ and the error term ε_i in (7.10) are statistically independent, it holds that

$$
\begin{aligned}
E\left[(y_i - \hat{y}_i)^2\right] &= E\left[(\varphi(\mathbf{x}_i) - \hat{y}_i)^2\right] + E\left[\varepsilon_i^2\right] \\
\Leftrightarrow \operatorname{var}\left[y_i - \hat{y}_i\right] &= E\left[\varphi(\mathbf{x}_i) - \hat{y}_i\right] + \operatorname{var}\left[\varepsilon_i^2\right]
\end{aligned}
\tag{7.11}
$$

which results in

$$
\sigma_p^2(\mathbf{x}) = \sigma_m^2(\mathbf{x}) + \sigma_\varepsilon^2
\tag{7.12}
$$

The assumption that the variance of the term σ_ε^2 is constant, has in practice proved simplistic, particularly in the case of financial problems. Hence, equation (7.3) can be rewritten using the following more realistic form:

$$
y_i = g_\lambda(\mathbf{x}_i; \mathbf{w}_0) + \varepsilon_i(\mathbf{x}_i)
\tag{7.13}
$$

In this case we have that

$$
\begin{aligned}
y_i - g_\lambda(\mathbf{x}_i; \hat{\mathbf{w}}_n) &= \{\varphi(\mathbf{x}_i) - g_\lambda(\mathbf{x}_i; \hat{\mathbf{w}}_n)\} + \varepsilon_i(\mathbf{x}_i) \\
\Rightarrow \operatorname{var}\left[y_i - \hat{y}_i\right] &= E\left[\varphi(\mathbf{x}_i) - \hat{y}_i\right] + \operatorname{var}\left[\varepsilon_i^2(\mathbf{x}_i)\right]
\end{aligned}
\tag{7.14}
$$

which will result in

$$\sigma_p^2(\mathbf{x}) = \sigma_m^2(\mathbf{x}) + \sigma_\varepsilon^2(\mathbf{x}) \tag{7.15}$$

CONSTRUCTING CONFIDENCE INTERVALS

To generate confidence intervals, distribution of the accuracy of the network prediction to the true underlying function is needed. In other words, the variance of the distribution of

$$f(\mathbf{x}) - \hat{y} \equiv f(\mathbf{x}) - g_\lambda(\mathbf{x}, \hat{\mathbf{w}}_n) \tag{7.16}$$

must be estimated. The model variance σ_m^2 will be estimated using two different bootstrapping methods: the bagging method, proposed by Breiman (1996), and the balancing method, proposed by Heskes (1997) and Carney et al. (1999). Both methods are variations of bootstrapping.

The Bagging Method

As mentioned earlier, ensemble network methods create different versions of the initial network and then combine the outputs to provide constancy to the predictor by stabilizing the high variance of the wavelet network. To do so, bootstrapping is used to create a new training sample from the initial data set. The algorithm of the bagging methods is described below.

In the first step, $B = 200$ new random samples with replacement are created from the original training sample. A new wavelet network is trained in each of the bootstrapped samples, $g_\lambda(\mathbf{x}^{(*i)}; \hat{\mathbf{w}}^{(*i)})$, where $(*i)$ indicates the ith bootstrapped sample and $\hat{\mathbf{w}}^{(*i)}$ is the solution of the ith bootstrapped sample. Each network is trained using the same topology as in the case of the original network (i.e., the same number of hidden units). Then each new network is evaluated to the original training sample, $\mathbf{x}$. In other words, we measure the forecasting accuracy of each bootstrapped network to the original data set.

The next step is to estimate the average output of the B networks using the original training sample $\mathbf{x}$:

$$g_{\lambda,\text{avg}}(\mathbf{x}) = \frac{1}{B} \sum_{i=1}^{B} g_\lambda(\mathbf{x}; \hat{\mathbf{w}}^{(*i)}) \tag{7.17}$$

It is assumed that the wavelet network produces an unbiased estimate of the underlying function $f(\mathbf{x})$. This means that the distribution of $P(f(\mathbf{x}) \mid g_{\lambda,\text{avg}}(\mathbf{x}))$ is centered on the estimate $g_{\lambda,\text{avg}}(\mathbf{x})$ (Carney et al., 1999; Heskes, 1997; Zapranis and Livanis, 2005). Since the wavelet network is not an unbiased estimator (as any other model), it is assumed that the bias component arising from the wavelet network is negligible compared to the variance component (Carney et al., 1999; Zapranis and Livanis, 2005). Finally, if we assume that the distribution of $P(f(\mathbf{x}) \mid g_{\lambda,\text{avg}}(\mathbf{x}))$ is normal, the model variance can be estimated by

$$\hat{\sigma}_m^2(\mathbf{x}) = \frac{1}{B-1} \sum_{i=1}^{B} \left[g_\lambda(\mathbf{x}; \hat{\mathbf{w}}^{(*i)}) - g_{\lambda,\text{avg}}(\mathbf{x}) \right]^2 \qquad (7.18)$$

To construct confidence intervals, the distribution of $P(g_{\lambda,\text{avg}}(\mathbf{x}) \mid f(\mathbf{x}))$ is needed. Since the distribution of $P(f(\mathbf{x}) \mid g_{\lambda,\text{avg}}(\mathbf{x}))$ is assumed to be normal, the "inverse" distribution $P(g_{\lambda,\text{avg}}(\mathbf{x}) \mid f(\mathbf{x}))$ is also normal. However, this distribution is unknown. Alternatively, it is estimated empirically by the distribution of $P(g_\lambda(\mathbf{x}) \mid g_{\lambda,\text{avg}}(\mathbf{x}))$ (Carney et al., 1999; Zapranis and Livanis, 2005). Then the confidence intervals are given by

$$g_{\lambda,\text{avg}}(\mathbf{x}) - t_{\alpha/2}\hat{\sigma}_m(\mathbf{x}) \le f(\mathbf{x}) \le g_{\lambda,\text{avg}}(\mathbf{x}) + t_{\alpha/2}\hat{\sigma}_m(\mathbf{x}) \qquad (7.19)$$

where $t_{\alpha/2}$ can be found in a Student's t table and $1 - a$ is the confidence level desired.

The Balancing Method

However, the estimator of the model variance, $\hat{\sigma}_m^2$, given by (7.18) is known to be biased (Carney et al., 1999); as a result, wider confidence intervals will be produced. Carney et al. (1999) proposed a balancing method to improve the model variance estimator. The algorithm for the balancing method is described below.

In the first strep, $B = 200$ new random samples with replacement are created from the original training sample. As in the bagging method, a new wavelet network is trained in each of the bootstrapped samples, $g_\lambda(\mathbf{x}^{(*i)}; \hat{\mathbf{w}}^{(*i)})$, where $(*i)$ indicates the ith bootstrapped sample and $\hat{\mathbf{w}}^{(*i)}$ is the solution of the ith bootstrapped sample. Each network is trained using the same topology as in the case of the original network. Then each new network is evaluated to the original training sample $\mathbf{x}$. In other words, we measure the forecasting accuracy of each bootstrapped network for the original data set.

Then the B bootstrapped samples are divided into M groups. More precisely, in our case the 200 ensemble samples are divided into $M = 8$ groups of 25 samples each. Next, the average output of each group is estimated:

$$\zeta = \left\{ g_{\lambda,\text{avg}}^{(i)}(\mathbf{x}) \right\}_{i=1}^{M} \qquad (7.20)$$

The model variance is not estimated just by the M ensemble output since this estimation will be highly volatile (Carney et al., 1999). To overcome this, a set of $P = 1000$ bootstraps of the values of ζ is created:

$$Y = \left\{ \zeta_j^* \right\}_{j=1}^{P} \qquad (7.21)$$

where

$$\zeta_j^* = \left\{ g_{\lambda,\text{avg}}^{(*j1)}(\mathbf{x}), g_{\lambda,\text{avg}}^{(*j2)}(\mathbf{x}), \ldots, g_{\lambda,\text{avg}}^{(*jM)}(\mathbf{x}) \right\} \tag{7.22}$$

is a bootstrapped sample of ζ. Then the model variance is estimated for each one of these sets by

$$\hat{\sigma}_j^{2*}(\mathbf{x}) = \frac{1}{M} \sum_{k=1}^{M} \left[g_{\lambda,\text{avg}}^{(*jk)}(\mathbf{x}) - g_{\lambda,\text{avg}}^{j}(\mathbf{x}) \right]^2 \tag{7.23}$$

where

$$g_{\lambda,\text{avg}}^{j}(\mathbf{x}) = \frac{1}{M} \sum_{k=1}^{M} g_{\lambda,\text{avg}}^{(*jk)}(\mathbf{x}) \tag{7.24}$$

Finally, the average model variance is estimated by taking the average of all $\hat{\sigma}_j^{2*}(\mathbf{x})$:

$$\hat{\sigma}_m^2(\mathbf{x}) = \frac{1}{P} \sum_{j=1}^{P} \hat{\sigma}_j^{2*}(\mathbf{x}) \tag{7.25}$$

This procedure is not computationally expensive since there is no need to train new networks. Hence, the complexity of both methods is similar and depends on the number B of the wavelet networks that must be trained.

Following the same assumptions as in the bagging method, confidence intervals can be constructed. Since a good estimator of the model variance is obtained, the improved confidence intervals using the balancing methods are given by

$$g_{\lambda,\text{avg}}(\mathbf{x}) - z_{\alpha/2}\hat{\sigma}_m(\mathbf{x}) \le f(\mathbf{x}) \le g_{\lambda,\text{avg}}(\mathbf{x}) + z_{\alpha/2}\hat{\sigma}_m(\mathbf{x}) \tag{7.26}$$

where $z_{\alpha/2}$ can be found in a standard Gaussian distribution table and $1 - a$ is the confidence level desired.

CONSTRUCTING PREDICTION INTERVALS

To generate prediction intervals, the distribution of the accuracy of the network prediction to target values is needed. In other words, the variance of the distribution of

$$y - \hat{y} \equiv y - g_\lambda\left(\mathbf{x}, \hat{\mathbf{w}}_n\right) \tag{7.27}$$

must be estimated. To construct prediction intervals, the total variance of the prediction, σ_p^2, must be estimated. As presented earlier, the total variance of the prediction is the sum of the model variance and the data noise variance. In the preceding section a method for estimating the model variance was presented. Here we emphasize a method for estimating the data noise variance.

The Bagging Method

First, the algorithm for creating predicting intervals using the bagging method is presented. To estimate the noise variance σ_ε^2, maximum likelihood methods are used. More precisely, a wavelet network will be trained on the residuals. An analytical description of the algorithm follows.

First, the initial wavelet network $g_\lambda(\mathbf{x}; \hat{\mathbf{w}}_n)$ is estimated and the solution $\hat{\mathbf{w}}_n$ of the loss function is found. Since it is assumed that the estimated wavelet network is a good approximation of the unknown underlying function, the vector $\hat{\mathbf{w}}_n$ is expected to be very close to the true vector $\mathbf{w}_0$ that minimizes the loss function.

In the next step the residuals between the network output and the target values are estimated. The noise variance can be approximated by a second wavelet network, $f_v(\mathbf{x}; \hat{\mathbf{u}}_n)$, where the squared residuals of the initial wavelet network are used as target values (Satchwell, 1994). In the second wavelet network, $f_v(\mathbf{x}; \hat{\mathbf{u}}_n)$, v is the number of hidden units and $\hat{\mathbf{u}}_n$ is the estimated vector of parameters that minimizes the loss function of the second wavelet network. Since it is assumed that the estimated wavelet network is a good approximation of the unknown underlying function, the vector $\hat{\mathbf{u}}_n$ is expected to be very close to the true vector $\mathbf{u}_0$ that minimizes the loss function. The following cost function is minimized in the second network:

$$\sum_{i=1}^{n} \left\{ [g_\lambda(\mathbf{x}_i; \mathbf{w}_0) - y_i]^2 - f_v(\mathbf{x}_i; \mathbf{u}_0) \right\}^2 \tag{7.28}$$

and for a new set of observations, $\mathbf{x}^*$, that were not used in the training, we have that

$$\hat{\sigma}_\varepsilon^2(\mathbf{x}^*) \approx f_v(\mathbf{x}^*; \mathbf{u}_0) \tag{7.29}$$

This technique assumes that the residual errors are caused by variance alone (Carney et al., 1999). To estimate the noise variance, data that were not used in the training of the bootstrapped sample should be used. One way to do this is to divide the data set in a training and a validation set. However, leaving out these test patterns is a waste of data (Heskes, 1997). Alternatively, an unbiased estimation of the output of the wavelet network, $\hat{y}_{ub}(\mathbf{x})$, can be approximated by

$$\hat{y}_{ub}(\mathbf{x}) = \frac{\sum_{i=i}^{B} q_i^m \hat{y}_i(\mathbf{x})}{\sum_{i=i}^{B} q_i^m} \tag{7.30}$$

where q_i^m is zero if the pattern m appears on the ith bootstrap sample and 1 otherwise. Constructing the new network $f_v(\mathbf{x}; \mathbf{u})$, we face the problem of model selection again. Using the methodology described in the preceding section, the correct number of v hidden units is selected. Usually, 1 or 2 hidden units are sufficient to model the residuals.

Using the estimation of the noise variance, the total variance can be calculated. The prediction intervals can be constructed using the following relationship:

$$g_{\lambda,\mathrm{avg}}(\mathbf{x}^*) - t_{\alpha/2}\hat{\sigma}_p(\mathbf{x}^*) \leq f(\mathbf{x}^*) \leq g_{\lambda,\mathrm{avg}}(\mathbf{x}^*) + t_{\alpha/2}\hat{\sigma}_p(\mathbf{x}^*) \qquad (7.31)$$

where $t_{\alpha/2}$ can be found in a Student's t distribution table and $1 - a$ is the confidence level desired.

The Balancing Method

As in the case of confidence intervals, the balancing method can be used to improve the accuracy of the intervals. The algorithm for estimating the noise variance is the same as in the case of the bagging method. However, the difference lies in the estimation of the model variance. Hence, if the balancing method is used, the prediction intervals are given by

$$g_{\lambda,\mathrm{avg}}(\mathbf{x}^*) - z_{\alpha/2}\hat{\sigma}_p(\mathbf{x}^*) \leq f(\mathbf{x}^*) \leq g_{\lambda,\mathrm{avg}}(\mathbf{x}^*) + z_{\alpha/2}\hat{\sigma}_p(\mathbf{x}^*) \qquad (7.32)$$

where $z_{\alpha/2}$ can be found in a standard Gaussian distribution table and $1 - a$ is the confidence level desired.

EVALUATING THE METHODS FOR CONSTRUCTING CONFIDENCE AND PREDICTION INTERVALS

In this section the bagging and balancing methods are evaluated in constructing confidence and prediction intervals. The two methods will be tested in the two functions $f(x)$ and $g(x)$, given by (3.15) and (3.17), respectively.

The confidence intervals are presented for the first function in Figure 7.3. The first part of the figure presents the confidence intervals using the bagging method, and the

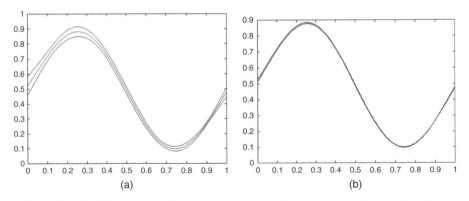

Figure 7.3   *Confidence intervals for the first case using the bagging (a) and balancing (b) methods.*

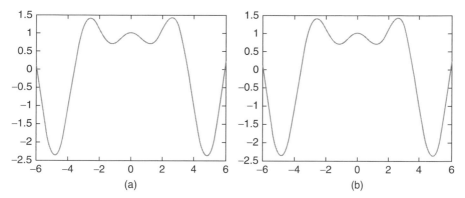

Figure 7.4   *Confidence intervals for the second case using the bagging (a) and balancing (b) methods.*

second part presents the confidence intervals using the balancing method. Similarly, Figure 7.4 presents the confidence intervals for the second function, where the first part refers to the bagging method, and the second part refers to the balancing method. It is clear that the confidence intervals using the balancing method are significantly narrower. This is due to the biased model variance estimator of the bagging method, which results in overestimation of the confidence intervals (Carney et al., 1999).

The 95% prediction intervals of the first function, $f(x)$, are presented in Figure 7.5. Again, the first part refers to the bagging method, and the second part refers to the balancing method. It is clear that both methods were able to capture the change in the variance of the noise. In both cases a wavelet network with 2 hidden units was used to approximate function $f(x)$, and a wavelet network with 1 hidden unit was used to approximate the residuals in order to estimate the noise variance. To compare the two methods, the prediction interval correct percentage (PICP) is used. PICP is the percentage of data points contained in the prediction intervals. Since the 95%

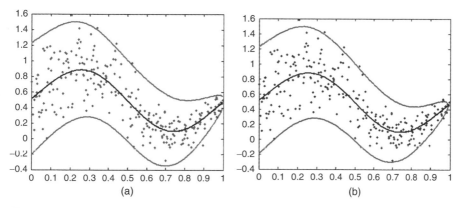

Figure 7.5   *Prediction intervals for the first case using the (a) bagging (PICP = 98%) and (b) balancing (PICP = 95%) methods.*

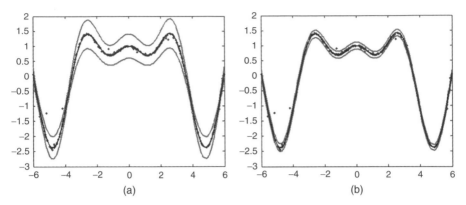

Figure 7.6 *Prediction intervals for the second case using the (a) bagging (PICP = 98.33%) and (b) balancing (PICP = 97.33%) methods.*

prediction intervals were estimated, a value of PICP close to 95 is expected. The bagging prediction intervals contain 98% of the data points (PICP), while in the case of the balancing method the PICP = 95% and is equal to the nominal value of 95%.

The same analysis is then repeated for the second function, $g(x)$. The 95% prediction intervals of $g(x)$ are presented in Figure 7.6. The first part refers to the bagging method and the second part refers to the balancing method. In both cases a wavelet network with 8 hidden units was used to approximate function $g(x)$, and a wavelet network with 2 hidden units was used to approximate the residuals in order to estimate the noise variance. As in the preceding case, the two methods are compared using the PICP. For the bagging method the PICP = 98.33%, while for the balancing method the PICP = 97.33%.

It is clear that the balancing method produced an improved estimator of the model variance. Our results are consistent with those of Breiman (1996), Carney et al. (1999), Heskes (1997), Papadopoulos et al. (2000), Zapranis and Livanis (2005), and Ζαπράνης (1999). In all cases the intervals produced by the balancing method were significantly smaller, while the PICP was considerable improved and closer to its nominal value.

CONCLUSIONS

In this chapter we described the main methodologies mentioned in the literature to estimate confidence intervals and prediction intervals for nonlinear nonparametric wavelet neural networks. In real applications, researchers and practitioners are generally more interested in prediction intervals. This is due to the fact that the prediction intervals are associated with the accuracy with which we can predict future prices and not just limited to the assessment of the correctness of the estimation of the actual function.

The model variance and the data noise variance are assumed to be independent. The total variance of the prediction is given by the sum of two variances. We assumed that the variance is not constant but changes over time.

Although the maximum likelihood method is considered biased, it can be used to estimate the variance, which depends on the network's input vector. Methods of random repetitive sampling, such as bootstrapping, make no assumptions about the nature of the noise; they can only estimate the uncertainty of the model variance and thus cannot be used to construct predictive intervals but only confidence intervals. For the construction of prediction intervals, an estimation of noise variance is needed. To approximate the noise variance a second wavelet network is trained, with the squared residuals of the initial wavelet network used as the target values.

We described two methods for constructing prediction and confidence intervals: bagging and balancing. Our results indicate that the balancing methods provide narrower confidence intervals and prediction intervals that produce more accurate PICP. The disadvantage of the iterative methodologies is that they are computationally expensive. In any case, analysis of the various approaches for estimating confidence and prediction intervals is an extremely interesting field of research.

REFERENCES

Breiman, I. (1996). "Bagging predictors." *Machine Learning*, 24, 123–140.

Carney, J. G., Cunningham, P., and Bhagwan, U. (1999). "Confidence and prediction intervals for neural network ensembles." *IJCNN'99*, Washington, DC.

De Veaux, D. R., Schumi, J., Schweinsberg, J., and Ungar, H. L. (1998). "Prediction intervals for neural networks via nonlinear regression." *Technometrics*, 40(4), 273–282.

Heskes, T. (1997). "Practical confidence and prediction intervals." *Advances in Neural Information Processing Systems*, 9, 176–182.

Papadopoulos, G., Edwards, J. P., and Murray, F. A. (2000). "Confidence estimation methods for neural networks: a practical comparison." *ESANN*, Bruges, Belgium, 75–80.

Satchwell, C. (1994). "Neural networks for stochastic problems: more than one outcome for the input space." *Neural Computing Applications Forum Conference*, Aston University, Birmingham, UK.

Zapranis, A. D., and Livanis, E. (2005). "Prediction intervals for neural network models." *Proceedings of the 9th WSEAS International Conference on Computers*, World Scientific and Engineering Academy and Society, Athens, Greece, 1–7.

Zapranis, A., and Refenes, A. P. (1999). *Principles of Neural Model Indentification, Selection and Adequacy: With Applications to Financial Econometrics*. Springer-Verlag, New York.

Ζαπράνης, Α. (2005). *Χρηματοοικονομική και Νευρωνικά Συστήματα*, Κλειδάριθμος, Αθήνα.

8

Modeling Financial Temperature Derivatives

In this chapter a real data set is used to demonstrate the application of our proposed framework. We focus on a financial application in forecasting the prices of weather derivatives. A weather derivative is a financial instrument that has a payoff derived from variables such as temperature, snowfall, humidity, and rainfall. Since their inception in 1996, weather derivatives have shown substantial growth. The first parties to arrange for, and issue, weather derivatives in 1996 were energy companies, which, after the deregulation of energy markets, were exposed to weather risk. In September 1999, the Chicago Mercantile Exchange (CME) launched the first exchange-traded weather derivatives. Today, weather derivatives are being used for hedging purposes by companies and industries whose profits can be affected adversely by unseasonal weather or, for speculative purposes, by hedge funds and others interested in capitalizing on those volatile markets.

Weather risk is unique in that it is highly localized, and despite great advances in meteorological science, still cannot be predicted precisely and consistently. Weather derivatives are also different from other financial derivatives in that the underlying weather index (i.e., HDD, CDD, CAT, etc.) cannot be traded. Furthermore, the corresponding market is relatively illiquid. Consequently, since weather derivatives cannot be replicated cost-efficiently by other weather derivatives, arbitrage pricing cannot apply to them directly. The weather derivatives market is a classically incomplete market, because the underlying weather variables are not tradable. When the market is incomplete, prices cannot be derived from the no-arbitrage condition, since it is

Wavelet Neural Networks: With Applications in Financial Engineering, Chaos, and Classification,
First Edition. Antonios K. Alexandridis and Achilleas D. Zapranis.

not possible to replicate the payoff of a given contingent claim by a controlled port-folio of basic securities. Consequently, the classical Black–Scholes–Merton pricing approach, which is based on no-arbitrage arguments, cannot be applied directly. In addition, market incompleteness is not the only reason; weather indices do not fol-low random walks (as the Black and Scholes approach assumes), and the payoffs of weather derivatives are determined by indices that are average quantities, whereas the Black–Scholes payoff is determined by the underlying value at exactly the maturity date of the contract (European options).

In the remainder of the chapter we introduce the weather derivative market and then focus on the modeling of the temperature process and the forecasting of future outcomes of weather derivatives. More precisely, using data from detrended and deseasonalized daily average temperatures (DATs), a wavelet network is constructed, initialized, and trained. At the same time, the significant variables are selected, in this case the correct number of lags. Finally, the trained wavelet network will be used to predict the future evolution of the weather derivative and to construct confidence and prediction intervals.

WEATHER DERIVATIVES

Weather derivatives are financial instruments that can be used by organizations or individuals as part of a risk management strategy to reduce risk associated with adverse or unexpected weather conditions. Just as traditional contingent claims, whose payoffs depend on the price of some fundamental underlying, a weather deriva-tive has an underlying measure, such as rainfall, temperature, humidity, or snowfall. The difference from other derivatives is that the underlying asset has no value and cannot be stored or traded; at the same time, the weather should be quantified in order to be introduced in the weather derivative. To do so, temperature, rainfall, precipita-tion, or snowfall indices are introduced as underlying assets. However, it is estimated that over 90% of the weather derivatives now traded are based on temperature.

Today, weather derivatives are being used for hedging purposes by companies and industries whose profits can be affected adversely by unseasonal weather or, for speculative purposes, by hedge funds and others interested in capitalizing on those volatile markets. According to Hanley (1999) and Challis (1999), nearly $1 trillion of the U.S. economy is directly exposed to weather risk. It is estimated that nearly 30% of the U.S. economy and 70% of U.S. companies are affected by weather (CME, 2005). The electricity sector is especially sensitive to temperature. According to Li and Sailor (1995) and Sailor and Munoz (1997), temperature is the most significant weather factor explaining electricity and gas demand in the United States. The impact of temperature on both electricity demand and price has been considered in many papers, including those of Engle et al. (1992), Henley and Peirson (1998), and Peirson and Henley (1994). Unlike insurance- and catastrophe-linked instruments, which cover high-risk and low-probability events, weather derivatives shield revenues against low-risk and high-probability events (e.g., mild or cold winters).

Today, the weather market is one of the fastest-developing markets. In 2004, the notional value of Chicago Mercantile Exchange (CME) weather derivatives was

$2.2 billion and grew tenfold to $22 billion through September 2005, with open interest exceeding $300,000 and volume surpassing 630,000 contracts traded. However, the over-the-counter (OTC) market was still more active than the exchange, so the bid–ask spreads were quite large.

Pricing and Modeling Methods

Early methods such as the actuarial method or the historical burn analysis (HBA) were used to derive the price of a temperature derivative written on a temperature index without actually modeling the dynamics of the temperature. Both methods measure how a temperature derivative would perform the previous years. The average (discounted) payoff that was derived from previous years is considered to be the payoff of the derivative (Jewson et al., 2005).

Alternatively, one can model the corresponding index directly, a method known as *index modeling*: the heating degree day (HDD) index, the cooling degree day (CDD) index, the cumulative average temperature (CAT) index, the accumulated HDDs (AccHDDs) index, or the accumulated CDDs (AccCDDs) index. A different model must be developed for each index. In the literature, a few papers suggest that temperature index modeling (HDD or CDD Index) might be more appropriate (Davis, 2001; Dorfleitner and Wimmer, 2010; Geman and Leonardi, 2005; Jewson et al., 2005).

Another approach to estimating the temperature-driving process is to use models based on daily temperatures. Daily modeling can, in principle, lead to more accurate pricing than that from modeling temperature indices (Jewson et al., 2005), as a lot of information is lost due to existing boundaries in the calculation of temperature indices by a normal or lognormal process, such as HDD being bounded by zero. On the other hand, deriving an accurate model for the daily temperature is not a straightforward process. Temperatures observed show seasonality in the mean, variance, distribution, and autocorrelation, and there is evidence of long memory in the autocorrelation. The risk with daily modeling is that small misspecifications in the models can lead to large mispricing of the temperature contracts (Jewson et al., 2005).

In the literature, two methods have been proposed for the modeling of the DAT, the use of a discrete or a continuous process. Moreno (2000) argues against the use of continuous processes in temperature modeling based on the fact that the values of temperature are in discrete form; hence, a discrete process should be used directly. Caballero and Jewson (2002), Caballero et al. (2002), Campbell and Diebold (2005), Cao et al. (2004), Cao and Wei (1999, 2000, 2003, 2004), Carmona (1999), Franses et al. (2001), Jewson and Caballero (2003a,b), Moreno (2000), Roustant et al. (2003a,b), Svec and Stevenson (2007), Taylor and Buizza (2002, 2004), and Tol (1996) make use of a general ARMA framework.

On the other hand, a wide range of studies suggest a temperature diffusion stochastic differential equation: Alaton et al. (2002), Bellini (2005), Benth (2003), Benth and Saltyte-Benth (2005, 2007), Benth et al. (2007, 2008), Bhowan (2003), Brody et al. (2002), Dischel (1998a,b, 1999), Dornier and Queruel (2000), Geman and Leonardi (2005), Hamisultane (2006a,b, 2007, 2008), McIntyre and Doherty (1999),

Oetomo and Stevenson (2005), Richards et al. (2004), Schiller et al. (2008), Torro et al. (2003), Yoo (2003), Zapranis and Alexandridis (2006, 2007, 2008, 2009a,b), and Zapranis and Alexandridis (2011). The continuous processes used for modeling daily temperatures usually take a mean-reverting form, which has to be discretized to estimate its various parameters. Once the parameters of the process are estimated, one can then value any contingent claim by taking the expectation of the discounted future payoff. Given the complex form of the process and the path-dependent nature of most payoffs, the pricing expression usually does not have closed-form solutions. In that case, MonteCarlo (MC) simulations are used. This approach typically involves generating a large number of simulated scenarios of weather indices to determine the possible payoffs of the weather derivative. The fair price of the derivative is then the average of all simulated payoffs, appropriately discounted for the time value of money. The precision of the MC approach depends on the correct choice of the temperature process and the look-back period of available weather data.

DATA DESCRIPTION AND PREPROCESSING

For accurate pricing and efficient weather risk management, the weather data must be of both adequate amount and highly quality (Dunis and Karalis, 2003). The data corresponding to the DATs for Berlin were provided by the European Climate Assessment and Dataset (ECAD; http://eca.knmi.nl/). The weather index we are interested in is the DAT. In the ECAD the DAT is measured as the average of the daily maximum and minimum temperature and is measured in Celsius degrees (°C). European weather contracts traded on the CME use the same measurement for the temperature. The precision with which temperature is measured in the ECAD is 0.1°C.

The data set consists of 3650 values, corresponding to 10 years' (1991–2000) of detrended and deseasonalized DATs in Berlin. So that each year would have equal observations, February 29 was removed from the data.

DATA EXAMINATION

In this section the data of the DATs of Berlin are examined to build a mathematical model that describes the dynamics of the temperature. In Figure 8.1a the DATs for Berlin for the period 1/1/1991 to 12/31/2000 are presented. Closer inspection of the figure reveals a seasonal cycle of one year, as expected. Moreover, extreme values in summer and winter are evident. To obtain a better insight of the temperature dynamics, the descriptive statistics of the DATs are examined. The mean temperature is 10.1, while variation of the DAT is quite large, with a standard deviation of 7.91. The difference between the maximum and minimum temperatures is around 44°C. These results indicate that temperature is very volatile and it is expected to be difficult to model and predict it accurately.

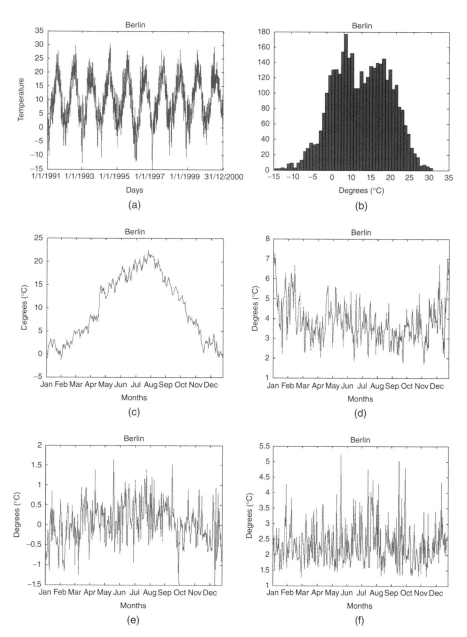

Figure 8.1 *(a) Daily average temperature, (b) empirical distribution, (c) mean, (d) standard deviation, (e) skewness, and (e) kurtosis of the DAT in Berlin.*

Negative skewness is evident, while the temperature in Berlin exhibits excess negative kurtosis. More precisely, the skewness is −0.08 and the kurtosis is 2.38. The results above indicate that the distribution of the DAT in Berlin is platykurtic, with a lower and wider peak where the mass of the distribution is concentrated on the left tail.

Finally, a normality test is performed. The normality is strongly rejected by a Jarque–Bera (JB) test. JB tests the null hypothesis that the sample data come from a normal distribution with unknown mean and variance, against the alternative that it does not come from a normal distribution:

$$H_0 = \text{sample comes from a normal distribution}$$
$$H_1 = \text{sample does not come from a normal distribution}$$

(8.1)

The JB test is a two-sided goodness-of-fit test suitable when a fully specified null distribution is unknown and its parameters must be estimated. The test statistic is given by

$$\text{JB} = \frac{n}{6}\left[s^2 + \frac{(k-3)^2}{4}\right]$$

(8.2)

where n is the sample size, s is the sample skewness, and k is the sample kurtosis. The JB statistic is over 62, and the p-value is zero, indicating the rejection of the null hypothesis that the temperature at the seven European cities follows a normal distribution. Figure 8.1b presents the empirical distribution of the DAT in Berlin. A closer inspection of the figure reveals a bimodal distribution, with the two peaks corresponding to the summer and winter temperatures.

To obtain better understanding of the temperature dynamics, the daily mean, standard deviation, skewness, and kurtosis of DAT were estimated. The mean of the DAT, $T_{\text{avg},t}$, was estimated using only observations for each particular day t. In Figure 8.1c the seasonal pattern is clear. The temperature has its highest values during the end of July and the beginning of August, while the lowest values are observed during the end of December and until the beginning of February. The mean DAT in Berlin fluctuates from −1.2 to 22.4°C.

Next, the standard deviation of the DAT is estimated. In Figure 8.1d the standard deviation for Berlin is presented. The standard deviation is higher in the winter period, while it is smaller in summer. Our results confirm the studies of Bellini (2005), Benth and Saltyte-Benth (2005, 2007), Benth et al. (2007), and Zapranis and Alexandridis (2008, 2009b).

Figure 8.1e presents the estimated skewness for each day t for Berlin. The figure reveals that the skewness tends to increase during the summer months while it decreases during the winter months. In general, the skewness is negative in winter months and positive in summer months. This means that it is more likely to have warmer days than average in summer and colder days than average in winter (Bellini, 2005).

Finally, the kurtosis on each day t can be found in Figure 8.1f. The figure does not reveal any seasonal pattern of the kurtosis. On the other hand, it is clear that the kurtosis has small deviations around 2, with many large upward spikes.

Two unit root tests were performed in the DAT. The temperature time series is tested for a unit root using an augmented Dickey–Fuller (ADF) test. The ADF was performed using the Schwartz information criterion in order to select the optimal number of lagged values. Our results reveal that the null hypothesis of a unit root is rejected since the ADF statistic is smaller than the critical value at the 5% significance level. More precisely, the ADF statistic is -5.1116, with p-value zero, while the critical value at the 5% significance level is -2.8621.

When a more powerful test is required, the Kwiatkowski et al. (1992) (KPSS) unit root test is employed. In contrast to the ADF test, the KPSS tests the null hypothesis that the time series is stationary versus the alternative that the time series is nonstationary (a unit root exists). The optimal bandwidth number was estimated using the Newery–West method. The KPSS statistic is 0.0505. The KPSS statistic has a value smaller than the critical value 0.463; hence, the null hypothesis that the time series is stationary cannot be rejected.

MODEL FOR THE DAILY AVERAGE TEMPERATURE: GAUSSIAN ORNSTEIN–UHLENBECK PROCESS WITH LAGS AND TIME-VARYING MEAN REVERSION

Many different models have been proposed to describe the dynamics of a temperature process. In this section a model for the seven cities studied in the preceding section is derived. Studying temperature data, Cao et al. (2004) and Cao and Wei (1999, 2000, 2003) built their framework on the following five assumptions about DAT:

- It follows a predicted cycle.
- It moves around a seasonal mean.
- It is affected by global warming and urban effects.
- It appears to have autoregressive changes.
- Its volatility is higher in winter than in summer.

As shown in the rest of the section, our results confirm the foregoing assumptions. It is known that temperature follows a predicted cycle. As expected and as shown in Figure 8.1, a strong cycle of one year is evident in all cities. It is also known that temperature has a mean-reverting form. Temperature moves around a seasonal mean and cannot deviate from that seasonal mean for long periods. This can be verified by Figure 8.1a,c, and d. In other words, it is not possible to observe temperatures of 20°C in winter in Oslo. Additionally, temperature is affected by global warming and urban effects. In areas under development, the surface temperature rises as more people and buildings concentrate. This is due to the sun's energy absorbed by the urban buildings and the emissions of vehicles, industrial buildings, and cooling units.

Hence, urbanization around a weather station results in an increment in the observed measurements of temperature. Finally, observing Figure 8.1d it is clear that the temperature volatility is higher in winter than in summer.

Following Zapranis and Alexandridis (2008, 2011) and Alexandridis (2010), a model that describes the temperature dynamics is given by a Gaussian mean-reverting Ornstein–Uhlenbeck process and is defined as follows:

$$dT(t) = dS(t) + \kappa(t)\,[T(t) - S(t)]\,dt + \sigma(t)\,dB(t) \tag{8.3}$$

where $T(t)$ is the average daily temperature, $\kappa(t)$ is the speed of mean reversion, $S(t)$ is a deterministic function modeling the trend and seasonality, $\sigma(t)$ is the daily volatility of temperature variations, and $B(t)$ is the driving noise process. As shown by Dornier and Queruel (2000), the term $dS(t)$ should be added for a proper mean-reversion toward the historical mean, $S(t)$.

Intuitively, it is expected that the speed of mean reversion is not constant. If the temperature today is away from the seasonal average (a cold day in summer), it is expected that the speed of mean reversion is high (i.e., the difference of today's and tomorrow's temperature is expected to be high). In contrast, if the temperature today is close to the seasonal variance, we expect the temperature to revert slowly to its seasonal average. To capture this feature, the speed of mean reversion is modeled by a time-varying function $\kappa(t)$.

It is clear that strong seasonalities and trends exist in the data. Before we present the time series to the wavelet network, both the trend and the seasonality must be removed. The usual approach is to model the seasonality by a sinusoid given by

$$S(t) = \text{Trend}_t + A\,\sin\frac{2\pi(t - f_i)}{365} \tag{8.4}$$

and the trend by a first-order polynomial given by

$$\text{Trend}_t = a + bt \tag{8.5}$$

Alexandridis and Zapranis (2013) used a more complex method where the seasonal mean and variance were extracted using wavelet analysis:

$$S(t) = \text{Trend}_t + \sum_{i=1}^{I_1} a_i \sin\frac{2\pi(t - f_i)}{p_i \cdot 365}$$

$$+ a_{I_1+1}\left[1 + \sin\frac{2\pi(t - f_{I_1+1})}{p_{I_1+1} \cdot 365}\right]\sin\frac{2\pi t}{365} \tag{8.6}$$

$$\sigma^2(t) = c + \sum_{i=1}^{I_2} c_i \sin\frac{2p_i'\pi t}{365} + \sum_{j=1}^{J_2} d_j \cos\frac{2p_j'\pi t}{365} \tag{8.7}$$

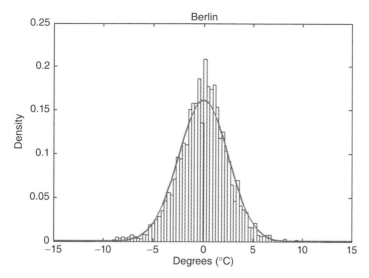

Figure 8.2   *Empirical and normal distribution (solid line) of the first difference of the daily average temperature in Berlin.*

To identify the terms I_1 and p_i in (8.6) and I_2, J_2, and p'_i in (8.7), we decompose the temperature series using a wavelet transform.

Finally, the driving noise process $B(t)$ is modeled by a standard Brownian motion (BM). The histogram of the first difference of the DAT and the normal distribution (solid line) is presented in Figure 8.2. A closer inspection of the figure reveals that the empirical distribution of the first difference of the DAT is similar to the normal distribution. Hence, selecting the BM as the driving noise process seems logical. This hypothesis is tested further later.

Following Benth and Saltyte-Benth (2007) and Zapranis and Alexandridis (2008), a discrete approximation of (8.3) is obtained and is given by

$$\Delta T(t) \approx \Delta S(t) + \kappa \left[T(t-1) - S(t-1) \right] + \sigma(t) \, \Delta B(t) \tag{8.8}$$

Next, by setting

$$\tilde{T}(t) = T(t) - S(t) \tag{8.9}$$

we have that

$$\tilde{T}(t) = a\tilde{T}(t-1) + \sigma(t)\varepsilon(t) \tag{8.10}$$

with

$$a = 1 + \kappa \tag{8.11}$$

Similarly, when the speed of mean reversion is a function of time, we have that

$$\tilde{T}(t) = a(t-1)\tilde{T}(t-1) + \sigma(t)\varepsilon(t) \tag{8.12}$$

$$a(t) = 1 + \kappa(t) \tag{8.13}$$

The detrended and deseasonalized temperature series, $\tilde{T}(t)$, can be modeled with an AR(1) process with a zero constant term, as shown in (8.12). In the context of such a model, the mean reversion parameter a is typically assumed to be constant over time. Brody et al. (2002) mentioned that in general a should be a function of time, but no evidence was presented. On the other hand, Benth and Saltyte-Benth (2005), using a data set comprising 10 years of Norwegian temperature data, calculated mean annual values of a. They reported that the variation of the values of a from year to year was not significant. They also investigated the seasonal structures in monthly averages of a and reported that none was found. However, to date no one has computed daily values of the mean reversion parameter, since there is no obvious way to do this in the context of model (8.12). On the other hand, averaging techniques, on a yearly or monthly basis, run the danger of filtering out too much variation and consequently of presenting a distorted picture regarding the true nature of a. The impact of a false specification of a on the accuracy of the pricing of temperature derivatives is significant (Alaton et al., 2002). However, Zapranis and Alexandridis (2008) estimated daily values of the variable $a(t)$ for the city of Paris using a nonparametric nonlinear NN. Their results indicate strong time dependence in the daily values of $a(t)$.

In this section we address that issue by using a wavelet network to estimate relationship (8.12) nonparametrically and then estimate a as a function of time. In addition, we apply the variable selection framework to select the number of lags in the temperature process. Hence, a series of speed of mean reversion parameters, $a_i(t)$, are estimated. By computing the derivative of the network output with respect to the network input, we obtain a series of daily values for $a_i(t)$. This gives us a much better insight into temperature dynamics and temperature derivative pricing. As shown by Alexandridis and Zapranis (2013), the daily variation of $a_i(t)$ is quite significant after all. In addition it is expected that the waveform of the wavelet network will provide a better fit to the DATs that are governed by seasonalities and periodicities.

In previous studies (Alaton et al., 2002; Bellini, 2005; Benth and Saltyte-Benth, 2005, 2007); Zapranis and Alexandridis, 2008, 2009b), it has been shown that an AR(1) model is not complex enough to completely remove the autocorrelation in the residuals. Alternatively, more complex models were suggested (Carmona, 1999; Geman and Leonardi, 2005). Using wavelet networks, the generalized version of (8.12) is estimated nonlinearly and nonparametrically; that is,

$$\tilde{T}(t+1) = \phi\left(\tilde{T}(t), \tilde{T}(t-1), ...\right) + e(t) \tag{8.14}$$

where

$$e(t) = \sigma(t)\varepsilon(t) \tag{8.15}$$

where $\tilde{T}$ is the detrended and deseasonalized DAT and the $e(t)$ are the residuals of the wavelet network.

After estimation of the wavelet network, the seasonal variance must be removed from the residuals. The exact modeling of the temperature process is beyond our scope in this chapter. To compare the out-of-sample forecast of a proposed model, the complete system, consisting of equations (8.5), (8.6), (8.7), (8.14), and (8.15), must be estimated. Here we focus on the construction of an optimal wavelet network for the system (8.14) that represents the detrended and deseasonalized DATs. Although the remaining steps are of no interest in this book, an out-of-sample comparison between the wavelet network and two linear models used widely in the literature is presented at the end of the chapter.

ESTIMATION USING WAVELET NETWORKS

Model (8.14) uses past temperatures (detrended and deseasonalized) over one period. Using more lags, we expect to overcome the strong correlation found in the residuals in models such as those of Alaton et al. (2002), Benth and Saltyte-Benth (2007), and Zapranis and Alexandridis (2008). However, the length of the lag series must be selected. Since the wavelet network is a nonparametrical nonlinear estimator, results from the autocorrelation function (ACF) or the partial ACF (PACF) cannot be used. Similarly, criteria used in linear models like the Schwarz criterion cannot be utilized. Hence, the variable significance algorithm presented in Chapter 5 is employed to determine the number of significant lags.

Variable Selection

The target values of the wavelet network are the detrended and deseasonalized DATs. The explanatory variables are lagged versions of the target variable. Choosing the length of a lag distribution in linear models can be done by minimizing an information criterion such as the Akaike or Schwarz criterion. Alternatively, the ACF and the PACF can be studied. The ACF suggests that the first 35 lags are significant. On the other hand, the PACF suggests that the first six lags, as well as lags 8 and 11, must be included in the model. However, results from these methods are not necessarily true in nonlinear nonparametric models.

Alternatively, to select only the significant lags, the variable selection algorithm presented in the preceding section will be employed. Initially, the training set contains the dependent variable and seven lags. Hence, the training set consists of 7 inputs, 1 output, and 3643 training pairs. The relevance of a variable to the model is quantified by the SBP criterion introduced earlier, which outperformed similar criteria.

Table 8.1 summarizes the results of the model identification algorithm for Berlin. Both the model selection and variable selection algorithms are included in the table. The algorithm concluded in four steps and the final model contains only three variables. The prediction risk for the reduced model is 3.1914, whereas for the original model it was 3.2004. On the other hand, the empirical loss increased slightly, from

TABLE 8.1　Variable Selection with Backward Elimination in Berlin[a]

Step	Variable to Remove (Lag)	Variable to Enter (Lag)	Variables in Model	Hidden Units (Parameters)	n/p Ratio	Empirical Loss	Prediction Risk
—	—	—	7	5 (83)	43.9	1.5928	3.2004
1	X_6	—	6	2 (33)	110.4	1.5922	3.1812
2	X_7	—	5	1 (17)	214.3	1.5927	3.1902
3	X_5	—	4	1 (14)	260.2	1.6004	3.2056
4	X_4	—	3	1 (11)	331.2	1.5969	3.1914

[a]The algorithm concluded in four steps. In each step the following are presented: which variable is removed, the number of hidden units for the particular set of input variables and the parameters used in the wavelet network, the empirical loss, and the prediction risk.

1.5928 for the initial model to 1.5969 for the reduced model, indicating that the explained variability (unadjusted) decreased slightly. However, the explained variability (adjusted for degrees of freedom) was increased for the reduced model to 64.61%, whereas it was 63.98% initially. Note that the noise term in $\tilde{T}$ is large, since the seasonal and trend parts were removed from the data; hence, we expect low values of R^2 and $\bar{R}^2$.

Finally, the number of parameters is reduced significantly in the final model. The initial model needed 5 hidden units and 7 inputs. Hence, 83 parameters were adjusted during the training phase, so the ratio of the number of training pairs n to the number of parameters p was 43.9. In the final model, only 1 hidden unit and 3 inputs were used, hence only 11 parameters were adjusted during the training phase and the ratio of the number of training pairs n to the number of parameters p was 331.2.

The statistics for the wavelet network model at each step are given in Table 8.2. More precisely, the first part of the table reports the value of the SBP and its p-value. In the second part of the table, various fitting criteria are reported: the mean absolute error, the maximum absolute error (MaxAE), the normalized mean squared error (NMSE), the mean absolute percentage error (MAPE), $\bar{R}^2$, the empirical loss, and the prediction risk.

In the full model it is clear that the value of the SBP for the last three variables is very small, in contrast to the first two variables. Observing the p-values, we conclude that the last four variables have p-value greater than 0.1, while the sixth lag has a p-value of 0.8826, strongly indicating a "not significant" variable. The wavelet network was converged after 43 iterations. In general, a very good fit was obtained. The empirical loss is 1.5928, and the prediction risk is 3.2004. MaxAE is 11.1823, MAE is 1.8080, and NMSE is 0.3521. MAPE is 3.7336. Finally, the $\bar{R}^2 = 63.98\%$.

The statistics for the wavelet network at step 1 are also presented in Table 8.2. The network had 6 inputs, two wavelets were used to construct the wavelet network, and 33 weights were adjusted during the training phase. The wavelet network converged after 17 iterations. By removing X_6 from the model, we observe from the table that the p-value of X_5 became 0, while for X_7 and X_4 the p-values became 0.5700 and 0.1403, respectively. The empirical loss was decreased slightly to 1.5922. However, MAE and NMSE were increased slightly, to 1.8085 and 0.3529, respectively. On the other hand, MaxAE and the MAPE were decreased to 11.1446 and 3.7127, respectively.

TABLE 8.2 Step-by-Step Variable Selection in Berlin[a]

Variable	Full Model SBP	Full Model p-Value	Step 1 SBP	Step 1 p-Value	Step 2 SBP	Step 2 p-Value	Step 3 SBP	Step 3 p-Value	Step 4 SBP	Step 4 p-Value
7	0.0026	0.7796	0.0031	**0.5700**						
6	0.0032	**0.8826**								
5	0.0053	0.6757	0.0131	0.0000	0.0206	**0.1907**				
4	0.0161	0.3500	0.0149	0.1403	0.0216	0.1493	−0.0052	**0.4701**		
3	0.2094	0.0000	0.2368	0.0000	0.2285	0.0000	0.1991	0.0000	0.2244	0.0000
2	1.1123	0.0000	1.0318	0.0000	1.0619	0.0000	0.9961	0.0000	0.9363	0.0000
1	9.8862	0.0000	10.0160	0.0000	9.9858	0.0000	10.0537	0.0000	10.1933	0.0000
MAE	1.8080		1.8085		1.8083		1.8093		1.8095	
MaxAE	11.1823		11.1446		11.1949		11.0800		11.0925	
NMSE	0.3521		0.3529		0.3525		0.3526		0.3530	
MAPE	3.7336		3.7127		3.7755		3.7348		3.7171	
R^2	63.98%		64.40%		64.59%		64.61%		64.61%	
Empirical loss	1.5928		1.5922		1.5927		1.6004		1.5969	
Prediction risk	3.2004		3.1812		3.1902		3.2056		3.1914	
Iterations	43		17		19		4		19	

[a]SBP was the average for each variable of 50 bootstrapped samples, the standard deviation, and the p-value; SBP, sensitivity-based pruning; MAE, mean absolute error; MaxAE, maximum absolute error; NMSE, normalized mean squared error; MSE, mean squared error; MAPE, mean absolute percentage error.

Next, the decision to remove X_6 is tested. The new prediction risk was reduced to 3.1812, while the explained variability adjusted for degrees of freedom increased to 64.40%. Hence, the removal of X_6 reduced the complexity of the model, while its predictive power was increased.

At step 2, X_7, which had the largest p-value, 0.5700, at the previous step, was removed from the model. Table 8.2 shows the statistics for the wavelet network at step 2. The new wavelet network had 5 inputs, 1 hidden unit was used, and 17 weights were adjusted during the training phase. The WN converged after 19 iterations. A closer inspection of Table 8.2 reveals that the removal of X_7 resulted in an increase in the error measures and a worse fit was obtained. The new $\bar{R}^2$ is 64.59%. The new prediction risk increased to 3.1902, which is smaller than the threshold. In other words, by removing X_7 the total predictive power of our model was slightly decreased; however, adding the variable X_7 to the model, only 0.28% additional variability of our model was explained, while the computational burden was increased significantly.

Examining the values of the SBP in Table 8.2 it is observed that the first two variables still have significantly larger values than the remaining variables. The p-values reveal that in the third step the X_5 must be removed from the model since its p-value is 0.1907.

At step 3 the network had 4 inputs, 1 hidden unit was used, and 14 weights were adjusted during the training phase. The wavelet network converged after four iterations. When removing X_5 from the model, we observe from Table 8.2 that only X_4 has a p-value greater than 0.1. Again the empirical loss and the prediction risk were increased. More precisely, the empirical loss is 1.6004 and the prediction risk increased from 0.48% to 3.2056%. The new prediction risk is greater than the estimated prediction risk of the initial model of about 0.16%. Again the increase in the prediction risk was significantly smaller than the threshold. On the other hand, $\bar{R}^2$ was increased to 64.61%, indicating an improved fit. Hence, the decision to remove X_5 was accepted.

In the final step, the variable X_4 had a p-value of 0.4701 and it was removed from the model. The network had 3 inputs, one wavelet was used in the construction of the wavelet network, and only 11 weights were adjusted during the training phase. The wavelet network converged after only 19 iterations. After the removal of X_4, a new wavelet network was trained with only one wavelet. The new empirical loss was decreased to 1.5969. The MAE and NMSE are 1.8095 and 0.3530, respectively, while MaxAE and MAPE are 11.0925 and 3.7171, respectively. Next the decision to remove X_4 was tested. The new prediction risk was reduced to 3.1914, while the explained variability adjusted for degrees of freedom was 64.61%. Hence, the removal of X_4 reduced the complexity of the model while its performance was increased. The p-values of the remaining variables are zero, indicating that the remaining variables are characterized as very significant variables. Hence, the algorithm stops. Our proposed algorithm indicates that only the three most recent lags should be used, while the PACF suggested the first six lags as well as lags 8 and 11.

In the final model only three of the seven variables were used. The complexity of the model was reduced significantly, since from 83 parameters in the initial model only 11 parameters have to be trained in the final model. In addition, in the reduced

TABLE 8.3   Prediction Risk at Each Step of the Variable Selection Algorithm for the First 5 Hidden Units for Berlin

Step	Hidden Units				
	1	2	3	4	5
0	3.2009	3.2026	3.2023	3.2019	**3.2004**
1	3.1817	**3.1812**	3.1828	3.1861	3.1860
2	**3.1902**	3.1915	3.1927	3.1972	3.1974
3	**3.2056**	3.2077	3.2082	3.2168	3.2190
4	**3.1914**	3.2020	3.2182	3.2158	3.2169

model the prediction risk was minimized when only 1 hidden unit was used, while 5 hidden units were needed initially. Our results indicate that the in-sample fit was slightly decreased in the reduced model. However, when an adjustment is made for the degrees of freedom, we observe that $\bar{R}^2$ was increased to 64.61% from 63.98% in the initial model. Finally, the prediction power of the final and less complex proposed model was improved since the prediction risk was reduced to 3.1914 from 3.2004.

Model Selection

In each step the appropriate number of hidden units is determined by applying the model selection algorithm presented in Chapter 4. Table 8.3 shows the prediction risk for the first 5 hidden units at each step of the variable selection algorithm for Berlin. Ideally, the prediction risk will decrease (almost) monotonically until a minimum is reached and will then start to increase (almost) monotonically. The number of hidden units that minimizes the prediction risk is selected for the construction of the model.

In the initial model, where all seven inputs were used, the prediction risk with 1 hidden unit is only 3.2009. When 1 additional hidden unit is added to the model, the prediction risk increases. Then, as more hidden units are added to the model, the prediction risk decreases monotonically. The minimum is reached when 5 hidden units are used and is 3.2004. When additional hidden units are added in the topology of the model, the prediction risk increases. Hence, the architecture of the wavelet network contains 5 hidden units. In other words, to construct the wavelet network, the five higher-ranking wavelets should be selected to form the wavelet basis. Observing Table 8.3, it is clear that the prediction risk in the initial model with only 1 hidden unit is almost the same as in the model with 5 hidden units. This is due to the small number of parameters that were adjusted during the training phase when only 1 hidden unit is used, not due to a better fit.

At the second step, when variable X_6 was removed, the prediction risk is minimized when 2 hidden units are used. Similarly, at steps 2, 3, and 4, the prediction risk is minimized when only 1 hidden unit is used. Additional hidden units do not improve the fitting or the predictive power of the model.

Initialization and Training

After the training set and the correct topology of the wavelet network are selected, the wavelet network can be constructed and trained. The BE method is used to initialize

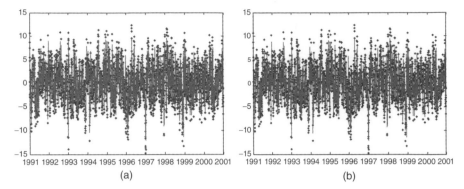

Figure 8.3 *Initialization of the final model for the temperature data in Berlin using the BE method (a) and the fit of the trained network with 1 hidden unit (b). The wavelet network converged after 19 iterations.*

the wavelet network. A wavelet basis is constructed by scanning the first four levels of the wavelet decomposition of the DAT in Berlin.

The wavelet basis consists of 168 wavelets. However, not all wavelets in the wavelet basis contribute to the approximation of the original time series. Following Zhang (1997), the wavelets that contain fewer than five sample points of the training data in their support are removed. Seventy-six wavelets that do not contribute significantly to the approximation of the original time series were identified. The truncated basis contains 92 wavelet candidates. Applying the BE method, the wavelets are ranked in order of significance. The wavelets in the wavelet library are ranked as follows: The BE starts the regression by selecting all the available wavelets from the wavelet library. Then the wavelet that contributes the least in the fitting of the training data is repeatedly eliminated. Since only 1 hidden unit is used on the architecture of the model, only the wavelet with the highest ranking is used to initialize the wavelet network. Figure 8.3a presents the initialization of the final model using only 1 hidden unit. The initialization is very good and the wavelet network converged after only 19 iterations. The training stopped when the minimum velocity, 10^{-5}, of the training algorithm was reached. The fitting of the trained wavelet network can be found in Figure 8.3b.

Next, various fitness criteria of the wavelet network corresponding to the $\tilde{T}_t$ in Berlin are estimated. Our results reveal that the wavelet networks fit the detrended and deseasonalized DATs reasonably well. The overall fit for Berlin is $\bar{R}^2 = 64.61\%$, while the MSE is 5.4196 and the MAE is only 1.8090. As mentioned earlier, the noise term is very large compared to the detrended and deseasonalized data, and as a result, a relatively small $\bar{R}^2$ is expected.

Next, the prediction of sign (POS) as well the prediction of change in direction (POCID) and the independent prediction of change in direction (IPOCID) are also estimated. These three criteria examine the ability of the network to predict changes, independent of the size of the change, and they are reported as percentages. The POS measures the ability of the network to predict the sign of the target values, positive or negative. The POS for the detrended and deseasonalized DATs is very high, 81.49%. The POCID is 60.15%, and the IPOCID is 52.30%.

For our temperature model we assumed that the noise term follows a normal distribution. Examining the residuals of the wavelet network, $e(t)$, given by equation (8.14), we find that the residuals have a mean of 0.01 and a standard deviation of 2.33. The skewness is −0.03 and the kurtosis is 3.73. Finally, the KS normality test has a p-value of 0 and rejects the normality of the residuals, while the Ljung–Box Q-statistic is 31.469 with a p-value of 0.0493, barely accepting the hypothesis of uncorrelated residuals.

However, after removal of the seasonal variance, the final residuals of the model, $\varepsilon(t)$, given by (8.15) have a mean of 0 and a standard deviation of 1. The skewness is −0.02 and the kurtosis is 3.53. Finally, the KS normality test has a p-value of 0.3081 and accepts the normality of the residuals, while the Ljung–Box Q-statistic is 29.616, with a p-value of 0.0763 accepting the hypothesis of uncorrelated residuals. Hence, our model does not violate our initial assumption and can be used for predictions.

Confidence and Prediction Intervals

After the wavelet network is constructed and trained, it can be used for prediction. Hence, confidence and prediction intervals can be constructed. In this section both confidence and prediction intervals are constructed using the balancing method. Using the BS method, 200 training samples are created and then divided into eight groups. In each group the average output of the wavelet networks is estimated. Next, new 1000 bootstrapped samples are created for the 8 average outputs to estimate the model variance given by (7.25). Then the confidence intervals is estimated with a level of significance a of 5%.

Figure 8.4 presents the confidence intervals for the detrended and deseasonalized DAT in Berlin as well as the average wavelet network output obtained from 200 bootstrapped samples. As the intervals are very narrow, to obtain a clear figure only the five first values are presented.

Next, the prediction intervals are constructed for the out-of-sample data set. The out-of-sample data consist of 365 values of detrended and deseasonalized DATs in Berlin for the period 2000–2001. The out-of-sample performance criteria are presented in Table 8.4. The overall fit adjusted for degrees of freedom is $\bar{R}^2 = 59.27\%$. The NMSE is 0.3961, while the MAPE is only 2.4108. Figure 8.5 illustrates the prediction intervals together with the real data and the average forecast of the wavelet network for the 200 bootstrapped samples. PICP = 93.46%.

Out-of-Sample Comparison

As mentioned earlier, to compare the out-of-sample forecast of a proposed model, the complete system consisting of equations (8.5), (8.6), (8.7), (8.14), and (8.15) must be estimated. An analytical discussion of temperature modeling in the context of weather derivatives was presented by Alexandridis and Zapranis (2013).

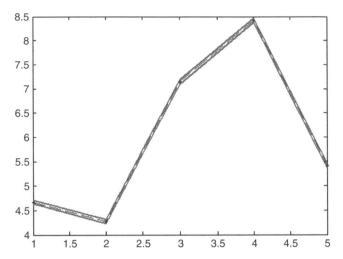

Figure 8.4 Confidence intervals and the average WN output using the balancing method and 200 bootstrapped samples. The five presents only the first five values for simplicity.

TABLE 8.4   Out-of-Sample Performance Criteria for Berlin

MAE	1.7340
MaxAE	9.3330
NMSE	0.3961
MAPE	2.4108
$\bar{R}^2$	59.27%

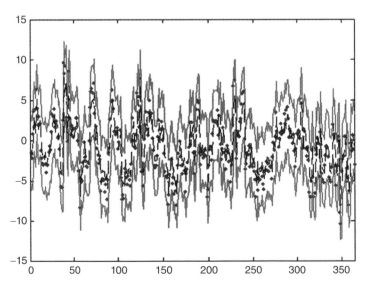

Figure 8.5 Prediction intervals (red solid lines), the real data (dotted), and the average wavelet network output (black solid line) using the balancing method and 200 bootstrapped samples of the detrended and deseasonalized DATs in Berlin for the period 2000–2001. PICP = 93.46%.

We presented only the part that is modeled by a wavelet network—in other words, the estimation of model (8.14)—but it would also be interesting to examine the forecasting ability of our model. To do so we examine the predictive power of the wavelet network against two linear models proposed previously and used widely in the literature: the first proposed by Alaton et al. (2002) and the second proposed by Benth and Saltyte-Benth (2007). The three models are similar, but in the two linear models, the speed of mean reversion, κ, is assumed to be constant and is computed through an AR(1) model given by (8.10).

The three models will be compared in the prediction of two temperature indices in seven European cities; the forecasting window will be 1, 2, 3, 6, and 12 months. Moreover, two forecasting methods will be used: one-day-ahead forecasting and out-of-sample forecasting. Hence, the three methods will be compared in 140 samples. The results indicate that the wavelet network outperforms the other two methods in 81 case (58%). On the other hand, the model proposed by Alaton et al. (2002) gave the best results in 35 cases (25%), while the model proposed by Benth and Saltyte-Benth (2007) gave the best results in only 24 cases (17%). For a complete analysis of temperature modeling and presentation of the forecasting comparison, we refer the reader to the book by Alexandridis and Zapranis (2013).

CONCLUSIONS

In this chapter a temperature time series was studied to develop a model that describes the temperature evolution in the context of weather derivative pricing. A mean reverting Ornstein–Uhlenbeck with seasonal mean and variance and time-varying speed of mean reversion was proposed.

In the context of an Ornstein–Uhlenbeck temperature process, the time dependence of the speed of the mean reversion $\kappa(t)$ was examined using a wavelet network. First, the variable selection framework proposed in Chapter 7 was utilized to estimate the number of lags of the nonlinear nonparametric AR model. At each step of the variable selection algorithm, the optimal number of wavelons was obtained by applying the minimum prediction risk criterion using bootstrapping.

The initial model had seven variables, whereas the final had one only three. Our results indicated that the reduced model has the same fitting to the data but increased predicting power. Moreover, a significantly smaller network and fewer parameters were needed during the training phase of the reduced wavelet network. Similarly, the training time was reduced significantly.

We compared the fit of the residuals with the normal distributions with two types of models. The first type was the nonlinear nonparametric model proposed, where κ is a function of time. The second category of models consists of two linear models proposed previously and often cited in the literature, where κ is constant. It follows that by setting the speed of mean reversion to be a function of time, the accuracy of the pricing of temperature derivatives improves. Generally, in our model a better fit was obtained.

Testing our model for adequacy, we find that the wavelet network does not violate any of the initial conditions. The residuals are uncorrelated and follow a normal distribution. On the other hand, in the two linear models, as presented by Alexandridis and Zapranis (2013), the normality hypothesis was rejected while strong autocorrelation in the residuals was evident.

The model proposed was evaluated out-of-sample. The predictive power of the model proposed was evaluated using two out-of-sample forecasting methods: out-of-sample forecasts over a period and one-day-ahead forecasts over a period. Modeling the DAT using wavelet networks significantly enhanced the fitting and the predictive accuracy of the temperature process. Modeling the DAT assuming a time-varying speed of mean reversion resulted in a model with better out-of-sample predictive accuracy. The additional accuracy of our model has an impact on the accurate pricing of temperature derivatives.

Concluding, wavelet networks can model very well the temperature process in the context of financial weather derivatives pricing. Consequently, they constitute an accurate and efficient tool for weather derivative pricing.

REFERENCES

Alaton, P., Djehince, B., and Stillberg, D. (2002). "On modelling and pricing weather derivatives." *Applied Mathematical Finance*, 9, 1–20.

Alexandridis, A. (2010). "Modelling and Pricing Temperature Derivatives Using Wavelet Networks and Wavelet Analysis," Ph.D. thesis, University of Macedonia, Thessaloniki, Greece.

Alexandridis, A. K., and Zapranis, A. D. (2013). *Weather Derivatives: Modeling and Pricing Weather-Related Risk*. Springer-Verlag, New York.

Bellini, F. (2005). "The Weather Derivatives Market: Modelling and Pricing Temperature," Ph.D. thesis, University of Lugano, Lugano, Switzerland.

Benth, F. E. (2003). "On arbitrage-free pricing of weather derivatives based on fractional Brownian motion." *Applied Mathematical Finance*, 10, 303–324.

Benth, F. E., and Saltyte-Benth, J. (2005). "Stochastic modelling of temperature variations with a view towards weather derivatives." *Applied Mathematical Finance*, 12(1), 53–85.

Benth, F. E., and Saltyte-Benth, J. (2007). "The volatility of temperature and pricing of weather derivatives." *Quantitative Finance*, 7(5), 553–561.

Benth, F. E., Saltyte-Benth, J., and Koekebakker, S. (2007). "Putting a price on temperature." *Scandinavian Journal of Statistics*, 34, 746–767.

Benth, F. E., Saltyte-Benth, J., and Koekebakker, S. (2008). *Stochastic Modelling of Electricity and Related Markets*. World Scientific, Singapore.

Bhowan, A. (2003). *Temperature Derivatives*. University of Witwatersrand, Johannesburg, South Africa.

Brody, C. D., Syroka, J., and Zervos, M. (2002). "Dynamical pricing of weather derivatives." *Quantitative Finance*, 2, 189–198.

Caballero, R., and Jewson, S. (2002). "Multivariate Long-Memory Modeling of Daily Surface Air Temperatures and the Valuation of Weather Derivative Portfolios." Working Paper. Retrieved July, 2002, from http://ssrn.com/abstract=405800

Caballero, R., Jewson, S., and Brix, A. (2002). "Long memory in surface air temperature: detection modelling and application to weather derivative valuation." *Climate Research*, 21, 127–140.

Campbell, S. D., and Diebold, F. X. (2005). "Weather forecasting for weather derivatives." *Journal of the American Statistical Association*, 100, 6–16.

Cao, M., and Wei, J. (1999). "Pricing Weather Derivatives: An Equilibrium Approach." Working Paper, Rotman Graduate School of Management, University of Toronto, Toronto.

Cao, M., and Wei, J. (2000). "Pricing the weather." Weather Risk Special Report, *Energy and Power Risk Management*, May, 67–70.

Cao, M., and Wei, J. (2003). "Weather Derivatives: A New Class of Financial Instruments." Working Paper, University of Toronto, Toronto, Canada.

Cao, M., and Wei, J. (2004). "Weather derivatives valuation and market price of weather risk." *Journal of Future Markets*, 24(11), 1065–1089.

Cao, M., Li, A., and Wei, J. (2004). "Watching the weather report." *Canadian Investment Review*, Summer, 27–33.

Carmona, R. (1999). "Calibrating degree day options." *3rd Seminar on Stochastic Analysis, Random Field and Applications*, Ecole Polytechnique de Lausanne, Ascona, Switzerland.

Challis, S. (1999). "Bright forecast for profits." *Reactions*, June.

CME. (2005). "An introduction to CME weather products." Retrieved January, 2007, from http://www.cme.com/edu/res/bro/cmeweather

Davis, M. (2001). "Pricing weather derivatives by marginal value." *Quantitative Finance*, 1, 1–4.

Dischel, B. (1998a). "At least: a model for weather risk." Weather Risk Special Report, *Energy and Power Risk Management*, March, 30–32.

Dischel, B. (1998b). "Black–Scholes won't do." Weather Risk Special Report, *Energy and Power Risk Management*, October, 8–9.

Dischel, B. (1999). "Shaping history for weather risk management." *Energy and Power Risk Management*, 12(8), 13–15.

Dorfleitner, G., and Wimmer, M. (2010). "The pricing of temperature futures at the Chicago Mercantile Exchange." *Journal of Banking and Finance*.

Dornier, F., and Queruel, M. (2000). "Caution to the wind." Weather Risk Special Report, *Energy and Power Risk Management*, August, 30–32.

Dunis, C. L., and Karalis, V. (2003). "Weather derivative pricing and filling analysis for missing temperature data." *Derivative Use, Trading and Regulation*, 9(1), 61–83.

Engle, R. F., Mustafa, C., and Rice, J. (1992). "Modelling peak electricity demand." *Journal of Forecasting*, 11, 241–251.

Franses, P. H., Neele, J., and van Dijk, D. (2001). "Modeling asymmetric volatility in weekly dutch temperature data." *Enivromental Modelling and Software*, 16, 37–46.

Geman, H., and Leonardi, M.-P. (2005). "Alternative approaches to weather derivatives pricing." *Managerial Finance*, 31(6), 46–72.

Hamisultane, H. (2006a). "Extracting Information from the Market to Price the Weather Derivatives." Working Paper. Retrieved November, 2006, from http://halshs.archives-ouvertes.fr/docs/00/17/91/89/PDF/weathderiv_extraction.pdf

Hamisultane, H. (2006b). "Pricing the Weather Derivatives in the Presence of Long Memory in Temperatures." Working Paper. Retrieved May, 2006, from http://halshs.archives-ouvertes.fr/docs/00/08/87/00/PDF/weathderiv_longmemory.pdf

Hamisultane, H. (2007). "Utility-Based Pricing of the Weather Derivatives." Working Paper. Retrieved September, 2006, from http://halshs.archives-ouvertes.fr/docs/00/17/91/88/PDF/weathderiv_utility.pdf

Hamisultane, H. (2008). "Which Method for Pricing Weather Derivatives?" Working Paper. Retrieved July, 2008, from http://halshs.archives-ouvertes.fr/docs/00/35/58/56/PDF/wpaper0801.pdf

Hanley, M. (1999). "Hedging the force of nature." *Risk Professional*, 1, 21–25.

Henley, A., and Peirson, J. (1998). "Residential energy demand and the interaction of price and temperature: British experimental evidence." *Energy Economics*, 20, 157–171.

Jewson, S., and Caballero, R. (2003a). "Seasonality in the dynamics of surface air temperature and the pricing of weather derivatives." *Meteorological Applications*, 10(4), 377–389.

Jewson, S., and Caballero, R. (2003b). "Seasonality in the statistics of surface air temperature and the pricing of weather derivatives." *Meteorological Applications*, 10(4), 367–376.

Jewson, S., Brix, A., and Ziehmann, C. (2005). *Weather Derivative Valuation: The Meteorological, Statistical, Financial and Mathematical Foundations*, Cambridge University Press, Cambridge, UK.

Kwiatkowski, D., Phillips, P. C. B., Schmidt, P., and Shin, Y. (1992). "Testing the null hypothesis of stationarity against the alternative of a unit root: How sure are we that economic time series have a unit root?" *Journal of Econometrics*, 54(1–3), 159–178.

Li, X., and Sailor, D. J. (1995). "Electricity use sensitivity and climate and climate change." *World Resource Review*, 3, 334–346.

McIntyre, R., and Doherty, S. (1999). "Weather risk: an example from the UK." *Energy and Power Risk Management*, June.

Moreno, M. (2000). "Riding the temp." *Weather Derivatives, FOW Special Supplement*, December.

Oetomo, T., and Stevenson, M. (2005). "Hot or cold? A comparison of different approaches to the pricing of weather derivatives." *Journal of Emerging Market Finance*, 4(2), 101–133.

Peirson, J., and Henley, A. (1994). "Electricity load and temperature issues in dynamic specification." *Energy Economic*, 16, 235–243.

Richards, T. J., Manfredo, M. R., and Sanders, D. R. (2004). "Pricing weather derivatives." *American Journal of Agricultural Economics*, 4(86), 1005–1017.

Roustant, O., Laurent, J.-P., Bay, X., and Carraro, L. (2003a). "A Bootstrap Approach to Price Uncertainty of Weather Derivatives." Ecole des Mines, Saint-Etienne and Ecole ISFA, Villeurbanne.

Roustant, O., Laurent, J.-P., Bay, X., and Carraro, L. (2003b). "Model Risk in the Pricing of Weather Derivatives." Ecole des Mines, Saint-Etienne and Ecole ISFA, Villeurbanne.

Sailor, D. J., and Munoz, R. (1997). "Sensitivity of electricity and natural gas consumption to climate in the USA: meteorology and results for eight states." *Energy, the international Journal*, 22, 987–998.

Schiller, F., Seidler, G., and Wimmer, M. (2008). "Temperature models for pricing weather derivatives." *Quantitative Finance*, 12(3), 489–500.

Svec, J., and Stevenson, M. (2007). "Modelling and forecasting temperature based weather derivatives." *Global Finance Journal*, 18(2), 185–204.

Taylor, J. W., and Buizza, R. (2002). "Neural network load forecasting with weather ensemble predictions." *IEEE Transactions on Power Systems*, 17(3), 626–632.

Taylor, J. W., and Buizza, R. (2004). "A comparison of temperature density forecasts from GARCH and atmospheric models." *Journal of Forecasting*, 23, 337–355.

Tol, R. S. J. (1996). "Autoregressive conditional heteroscedasticity in daily temperature measurements." *Environmetrics*, 7, 67–75.

Torro, H., Meneu, V., and Valor, E. (2003). "Single factor stochastic models with seasonality applied to underlying weather derivatives variables." *Journal of Risk Finance*, 4(4), 6–17.

Yoo, S. (2003). "Weather Derivatives and Seasonal Forecast." Working Paper, Department of Applied Economics and Management, Cornell University, Ithaca, NY.

Zapranis, A., and Alexandridis, A. (2006). "Wavelet Analysis and Weather Derivatives Pricing." 5th Hellenic Finance and Accounting Association, Thessaloniki, Greece.

Zapranis, A., and Alexandridis, A. (2007). "Weather Derivatives Pricing: Modelling the Seasonal Residuals Variance of an Ornstein–Uhlenbeck Temperature Process with Neural Networks." *EANN 2007*, Thessaloniki, Greece.

Zapranis, A., and Alexandridis, A. (2008). "Modelling temperature time dependent speed of mean reversion in the context of weather derivetive pricing." *Applied Mathematical Finance*, 15(4), 355–386.

Zapranis, A., and Alexandridis, A. (2009a). "Modeling and Forecasting CAT and HDD Indices for Weather Derivative Pricing." *EANN*, London,UK.

Zapranis, A., and Alexandridis, A. (2009b). "Weather derivatives pricing: modelling the seasonal residuals variance of an Ornstein–Uhlenbeck temperature process with neural networks." *Neurocomputing*, 73, 37–48.

Zapranis, A., and Alexandridis, A. (2011). "Modeling and forecasting cumulative average temperature and heating degree day indices for weather derivative pricing." *Neural Computing and Applications*, 20(6), 787–801.

Zhang, Q. (1997). "Using wavelet network in nonparametric estimation." *IEEE Transactions on Neural Networks*, 8(2), 227–236.

9

Modeling Financial Wind Derivatives

Weather derivatives are financial tools that can help organizations or individuals to reduce risk associated with adverse or unexpected weather conditions and can be used as part of a risk management strategy. Weather derivatives linked to various weather indices, such as rainfall, temperature, or wind, are traded extensively in CME as well as on the OTC market. The electricity sector is especially sensitive to the temperature and wind since temperature affects the consumption of electricity while wind affects production of electricity in wind farms. Hence, it is logical that energy companies are the main investors of the weather market. In Chapter 8 a detailed framework for modeling and pricing temperature derivatives was developed. In this chapter we focus on wind derivatives.

The notional value of the wind-linked securities traded is around \$36 million, indicating a large and growing market (WRMA, 2010). However, after the close of the U.S. Future Exchange, wind derivatives have been traded OTC. A demand for these derivatives exists. However, investors hesitate to enter into wind contracts. The main reasons for the slow growth of the wind market compared to temperature contracts are the difficulty in modeling wind accurately and the challenge of finding a reliable model for valuing related contracts. As a result, there is a lack of reliable valuation framework that makes financial institutions reluctant to quote prices over these derivatives.

The aim of this chapter is to model and price wind derivatives. Wind derivatives are standardized products that depend only on the daily average wind speed measured

Wavelet Neural Networks: With Applications in Financial Engineering, Chaos, and Classification, First Edition. Antonios K. Alexandridis and Achilleas D. Zapranis.
© 2014 John Wiley & Sons, Inc. Published 2014 by John Wiley & Sons, Inc.

by a predefined meteorological station over a specified period and can be used by wind (and weather in general)-sensitive businesses such as wind farms, transportation companies, construction companies, and theme parks. The financial contracts that are traded are based on the simple daily average wind speed index, and this is the reason that we choose to model only the dynamics of the daily average wind speeds. The revenues of each company have a unique dependence and sensitivity to wind speeds. Although wind derivatives and weather derivatives can hedge a significant part of the weather risk of the company, some basis risk will always still exist which must be hedged from each company separately. This can be done either by defining a more complex wind index or by taking an additional hedging position. Cao and Wei (2003) provided various examples of how weather derivatives can reduce the basis and volumetric risks in weather-sensitive businesses (e.g., ski resorts, restaurants, theme parks, electricity companies).

Wind is a free, renewable, and environmentally friendly source of energy (Billinton et al., 1996). Wind derivatives are traded extensively in the electricity sector. While the demand for electricity is closely related to the temperature, the electricity produced by a wind farm is dependent on the wind conditions. The risk exposure of the wind farm depends on the wind speed, the wind direction, and in some cases the wind duration of the wind speed at certain levels. However, modern wind turbines include mechanisms that allow turbines to rotate in the appropriate wind direction (Caporin and Pres, 2010). Wind derivatives can be used as part of a hedging-strategy in wind-sensitive businesses. However, the underlying wind indices do not account for the duration of the wind speed at certain levels but, rather, usually measure the average daily wind speed. Hence, the parameter of the duration of the wind speed at certain levels is not considered in our daily model. Hence, the risk exposure of a wind-sensitive company can be analyzed by quantifying only the wind speed. On the other hand, companies such as wind farms whose revenues depend on the duration effect can use an additional hedging strategy that includes this parameter. This can be done by introducing a second index that measures the duration. A similar index for temperature is the frost day index.

Many different approaches have been proposed so far for modeling the dynamics of the wind speed process. The most common is the generalized autoregressive moving average (ARMA) approach. There has been a number of studies on the use of linear ARMA models to simulate and forecast wind speed in various locations (Billinton et al., 1996; Caporin and Pres, 2010; Castino et al., 1998; Daniel and Chen, 1991; Huang and Chalabi, 1995; Kamal and Jafri, 1997; Martin et al., 1999; Saltyte-Benth and Benth, 2010; Tol, 1997; Torres et al., 2005). Kavasseri and Seetharaman (2009) used a more sophisticated fractional integrated ARMA (ARFIMA) model. Most of these studies did not consider in detail the accuracy of the wind speed forecasts (Huang and Chalabi, 1995). On the other hand, Ailliot et al. (2006) apply an autoregressive model (AR) with time-varying coefficients to describe the space-time evolution of wind fields. Benth and Saltyte-Benth (2009) introduced a stochastic process called the continuous AR (CAR) model to model and forecast daily wind speeds. Finally, Nielsen et al. (2006) presented various statistical methods for short-term wind speed forecasting. Sfetsos (2002) argues about the use of linear or meteorological models, since their prediction error is not significantly lower than the elementary persistent method. Alternatively, some studies use space-state models to fit the speed and the direction of the wind simultaneously (Castino et al., 1998; Cripps et al., 2005; Haslett and Raftery, 1989; Martin et al., 1999; Tolman and Booij, 1998; Tuller and Brett, 1984).

Alternative to the linear models, artificial intelligence was used for wind speed modeling and forecasting. Alexiadis et al. (1998), Barbounis et al. (2006), Beyer et al. (1994), Mohandes et al. (1998), More and Deo 2003, and Sfetsos (2000, 2002) utilized neural networks to model the dynamics of the wind speed process. Mohandes et al. (2004) used support vector machines, and Pinson and Kariniotakis (2003) employed fuzzy neural networks.

Depending on the application, wind modeling is based on an hourly (Ailliot et al., 2006; Castino et al., 1998; Daniel and Chen, 1991; Kamal and Jafri, 1997; Martin et al., 1999; Sfetsos, 2000, 2002; Torres et al., 2005; Yamada, 2008), daily (Benth and Saltyte-Benth, 2009; Billinton et al., 1996; Caporin and Pres, 2010; Huang and Chalabi, 1995; More and Deo, 2003; Tol, 1997), weekly, or monthly basis (More and Deo, 2003). When the objective is to hedge against electricity demand and production, hourly modeling is used, whereas for weather derivative pricing the daily method is used. More rarely, weekly or monthly modeling is used to estimate monthly wind indexes. Since we want to focus on weather derivative pricing, the daily modeling approach is followed; however, the method proposed can be easily be adapted to hourly modeling.

Wind speed modeling is much more complicated than temperature modeling since wind has a direction and is greatly affected by the surrounding terrain, such as buildings, and trees (Jewson et al., 2005). However, Benth and Saltyte-Benth (2009) have shown that wind speed dynamics share a lot of common characteristics with the dynamics of temperature derivatives. In this context we use a mean reverting Ornstein–Uhlenbeck stochastic process to model the dynamics of the wind speed, where the innovations are driven by a Brownian motion. The statistical analysis reveals seasonality in the mean and variance. In addition we use a novel approach to model the autocorrelation of wind speeds. More precisely, a wavelet network is utilized to capture accurately the autoregressive characteristics of the wind speeds.

The evaluation of the proposed methodology against alternative modeling proce-dures proposed in prior studies indicates that wavelet networks can accurately model and forecast the dynamics and evolution of the speed of the wind. The performance of each method was evaluated in-sample as well as out-of-sample and for different time periods.

The remainder of the chapter is organized as follows. First, a statistical analysis of the wind speed dynamics is presented. Then, a linear model and a nonlinear nonparametric wavelet network are fitted to the data. Next, an evaluation and a comparison of the models studied is presented. Finally, we conclude.

MODELING THE DAILY AVERAGE WIND SPEED

In this section we derive empirically the characteristics of the daily average wind speed (DAWS) dynamics in New York. The data were collected from NOAA

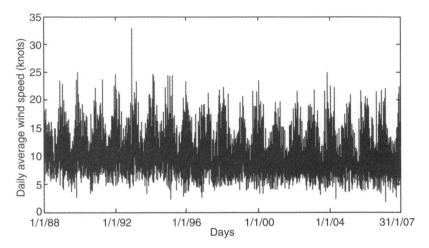

Figure 9.1 *Daily average wind speed for New York.*

(http://www.noaa.gov/) and correspond to DAWSs. The wind speeds are measured in units of 0.1 knot. The measurement period is between January 1, 1988 and February 28, 2008. The first 20 years are used for estimation of the parameters, while the remaining two months are used for evaluation of the performance of the model proposed. So that each year will have the same number of observations, February 29 is removed from the data, resulting in 7359 data points. The time series is complete without any missing values.

In Figure 9.1 the DAWSs for the first 20 years are presented. The descriptive statistics of the in-sample data are presented in Table 9.1. The values of the data are always positive and range from 1.8 to 32.8 with a mean around 9.91. Also, a closer inspection of Figure 9.1 reveals seasonality.

The descriptive statistics of the DAWSs indicate that there is a strong positive kurtosis and skewness, while the normality hypothesis is rejected based on the Jarque–Bera statistic. The same conclusions can be reached by observing the first part of Figure 9.2, where the histogram of the DAWSs is represented. The distribution of DAWSs deviates significantly from the normal and is not symmetrical. In the literature the Weibull or the Rayleigh (which is a special case of the Weibull) distributions were proposed describe the distribution of the wind speed (Brown et al., 1984; Daniel and Chen, 1991; Garcia et al., 1998; Justus et al., 1978; Kavak Akpinar and Akpinar, 2005; Nfaoui et al., 1996; Torres et al., 2005; Tuller and Brett, 1984). In addition, some

TABLE 9.1 **Descriptive Statistics of the Wind Speed in New York**[a]

	Mean	Med.	Max.	Min.	S. Dev.	Skew	Kurt.	JB	p-Value
Original	9.91	9.3	32.8	1.8	3.38	0.96	4.24	1595.41	0
Transformed	2.28	2.3	3.6	0.6	0.34	0.00	3.04	0.51	1

[a]JB, Jarque–Bera statistic; p-value, p-values of the JB statistic.

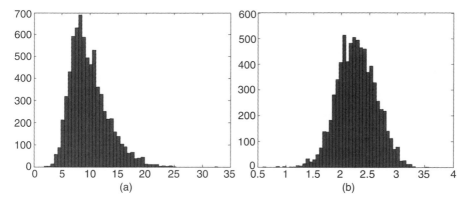

Figure 9.2   *Histogram of the (a) original and (b) Box–Cox transformed data.*

studies propose use of the lognormal distribution (Garcia et al., 1998) or chi-square (Dorvlo, 2002). Finally, Jaramillo and Borja (2004) used a bimodal Weibull and Weibull distribution. However, empirical studies favor use of the Weibull distribution (Celik 2004; Tuller and Brett, 1984).

A closer inspection of Figure 9.2a reveals that the DAWSs in New York follow a Weibull distribution with scale parameter $\lambda = 11.07$ and shape parameter $k = 3.04$. To symmetrize the data, the Box–Cox transform is applied. The Box–Cox transformation is given by

$$W^{(l)} = \begin{cases} \frac{W^l - 1}{l} & l \neq 0 \\ \ln(W) & l = 0 \end{cases} \tag{9.1}$$

where $W^{(l)}$ is the transformed data. The parameter l is estimated by maximizing the log-likelihood function. Note that log-transform is a special case of the Box–Cox transform with $l = 0$. The optimal l of the Box–Cox transform for the DAWSs in New York is estimated to be 0.014. In Figure 9.2b the histogram of the transformed data can be found while the second row of Table 9.1 shows the descriptive statistics of the transformed data.

The DAWSs exhibit a clear seasonal pattern which is preserved in the transformed data. The same conclusion can be reached by examining the autocorrelation function (ACF) of the DAWSs in the first part of Figure 9.3. The seasonal effects are modeled by a truncated Fourier series, given by

$$S(t) = a_0 + b_0 t + \sum_{i=1}^{I_1} a_i \sin \frac{2\pi i(t - f_i)}{365} + \sum_{j=1}^{J_1} b_j \sin \frac{2\pi j(t - g_j)}{365} \tag{9.2}$$

In addition, we examine the data for a linear trend representing the global warming or the urbanization around the meteorological station. First, we quantify the trend by fitting a linear regression to the DAWS data. The regression is statistically significant

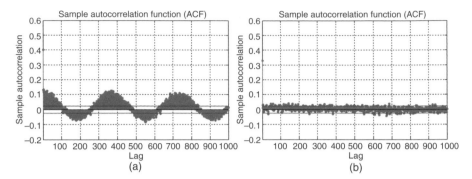

Figure 9.3 *Autocorrelation function of the transformed DAWSs in New York (a) before and (b) after removing the seasonal mean.*

TABLE 9.2   Estimated Parameters of the Seasonal Component

a_0	b_0	a_1	f_1	b_1	g_1
2.3632	−0.000024	0.0144	827.81	0.1537	28.9350

with intercept $a_0 = 2.3632$ and slope $b_0 = -0.000024$, indicating a slight decrease in the DAWSs. Next, the seasonal periodicities are removed from the detrended data. The remaining statistically significant estimated parameters of equation (9.2) with $I_1 = J_1 = 1$ are presented in Table 9.2. As shown in Figure 9.3b, the seasonal mean was removed successfully. The same conclusion was reached in previous studies for daily models for both temperature and wind (Benth and Saltyte-Benth, 2009; Zapranis and Alexandridis, 2008).

LINEAR ARMA MODEL

In the literature, various methods have been proposed for studying the statistical characteristics of the wind speed, in daily or hourly measurements. However, the majority of the studies utilize variations of the general ARMA model. In this chapter we first estimate the dynamics of the detrended and deseasonalized DAWS process using a general ARMA model and then compare our results with a wavelet network. We define the detrended and deseasonalized DAWS as

$$\tilde{W}^{(l)}(t) = W^{(l)}(t) - S(t) \tag{9.3}$$

The dynamics of $\tilde{\mathbf{W}}^{(l)}(t)$ are modeled by a vectorial Ornstein–Uhlenbeck stochastic process,

$$d\tilde{W}^{(l)}(t) = \mathbf{a}\tilde{W}^{(l)}(t)\,dt - I_p\sigma(t)\,dB_t \tag{9.4}$$

where I_p is the pth unit vector in R^p and $\sigma(t)$ is the standard deviation.

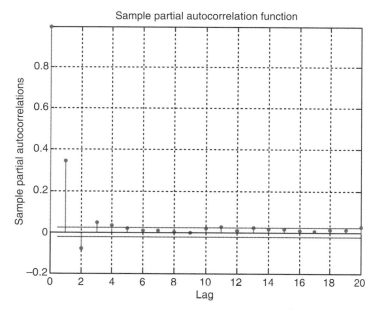

Figure 9.4   *Partial autocorrelation function of the detrended and deseasonalized DAWSs in New York.*

First, to select the correct ARMA model, we examine the ACF of the detrended and deseasonalized DAWSs. A closer inspection of Figure 9.3b reveals that the lags 1, 2, and 4 are significant. On the other hand, by examining the partial autocorrelation function (PACF) in Figure 9.4, we conclude that the first four lags are necessary to model the autoregressive effects of the dynamics of the wind speed.

To find the correct model, we estimate the log-likelihood function (LLF) and the akaike information criterion (AIC). Consistent with the PACF, both criteria suggest that an AR(4) model is adequate for modeling the wind process since they were minimized when a model with four lags was used. The estimated parameters and the corresponding p-values are presented in Table 9.3. It is clear that the three first parameters are statistically very significant since their p-value is less than 0.05. The parameter of the fourth lag is statistically significant with a p-value of 0.0657. The AIC for this model is 0.46852 and the LLF is -1705.14.

Observing the residuals of the AR model in Figure 9.5a, we conclude that the autocorrelation was removed successfully. However, the ACF of the squared residuals indicates a strong seasonal effect in the variance of the wind speed, as shown in Figure 9.6. Similar behavior was observed in the residuals of temperature and wind in

TABLE 9.3   Estimated Parameters of the Linear AR(4) Model

Parameter	AR(1)	AR(2)	AR(3)	AR(4)
Value	0.3617	−0.0999	0.0274	0.0216
p-Value	0.0000	0.0000	0.0279	0.0657

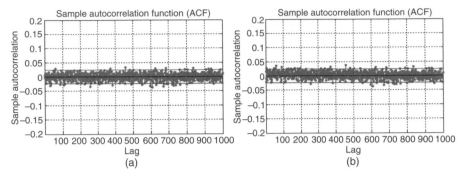

Figure 9.5   *Autocorrelation function of the residuals of (a) the linear model and (b) the WN.*

various studies (Zapranis and Alexandridis, 2008). The seasonal variance is modeled with a truncated Fourier series:

$$\sigma^2(t) = c_0 + \sum_{i=1}^{I_2} c_i \sin \frac{2i\pi t}{365} + \sum_{j=1}^{J_2} d_j \sin \frac{2\pi j t}{365} \qquad (9.5)$$

Note that we assume that the seasonal variance is periodic and repeated every year [i.e., $\sigma^2(t + 365) = \sigma^2(t)$, where $t = 1, \dots, 7359$]. The empirical and fitted seasonal variance are presented in Figure 9.7 and the estimated parameters of equation (9.5) are presented in Table 9.4.

Not surprisingly, the variance exhibits the same characteristics as in the case of temperature. More precisely, the seasonal variance is higher in the winter and early summer, while it reaches its lower values during the summer period.

Finally, the descriptive statistics of the final residuals are examined. A closer inspection of Table 9.5 shows that the autocorrelation has been removed successfully, as indicated by the Ljung–Box Q-statistic. In addition, the distribution of the residuals is very close to the normal distribution, as shown in Figure 9.8a. However, small

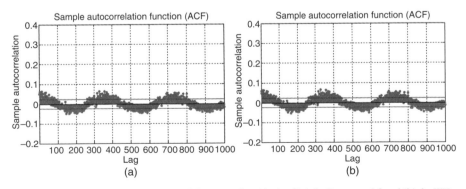

Figure 9.6   *Autocorrelation function of the squared residuals of (a) the linear model and (b) the WN.*

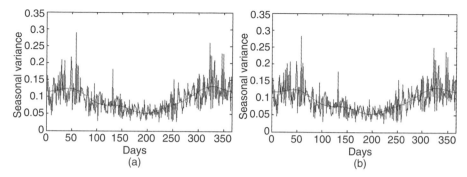

Figure 9.7 *Empirical and fitted seasonal variance of (a) the linear model and (b) the WN.*

TABLE 9.4   Estimated Parameters of the Seasonal Variance in the Linear Model

c_0	c_1	c_2	c_3	c_4	d_1	d_2	d_3	d_4
0.0932	0.000032	−0.0041	0.0015	−0.0028	0.0358	−0.0025	−0.0048	−0.0054

TABLE 9.5   Descriptive Statistics of the Residuals for the Linear AR(4) Model

Var.	Mean	S. dev.	Max.	Min.	Skew	Kur.	JB	*p*-Value	KS	*p*-Value	LBQ	*p*-Value
Noise	0	1	3.32	−5.03	−0.09	3.03	10.097	0.007	1.033	0.2349	8.383	0.989

S. dev., standard deviation; JB, Jarque–Bera statistic; KS, Komogorov–Smirnov statistic; LBQ, Ljung–Box Q–statistic.

negative skewness exists. More precisely, the residuals have mean 0 and standard deviation 1. In addition, the kurtosis is 3.03 and the skewness is −0.09.

Concluding, the previous analysis indicates that an AR(4) model provides a good fit for the wind process, while the final residuals are very close to the normal distribution.

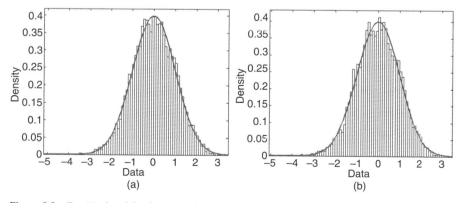

Figure 9.8 *Empirical and fitted normal distribution of the final residuals of (a) the linear model and (b) the WN.*

WAVELET NETWORKS FOR WIND SPEED MODELING

In this section, wavelet networks are applied in the transformed, detrended, and deseasonalized wind speed data in order to model the daily dynamics of wind speeds in New York. Motivated by the waveform of the data, we expect a wavelet function to better fit the wind speed. In addition, it is expected that the nonlinear form of the wavelet network will provide more accurate representation of the dynamics of the wind speed process both in-sample and out-of-sample.

In the context of the linear ARMA model, the mean reversion parameter a is typically assumed to be constant over time. Brody et al. (2002) mentioned that, in general, a should be a function of time, but no evidence was presented. The impact of a false specification of a on the accuracy of the pricing of temperature derivatives is significant (Alaton et al., 2002).

In this section we address that issue by using a wavelet network to estimate nonparametrically the relationship (9.4) and then estimate **a** as a function of time. In addition, we propose the variable selection algorithm presented in previous chapters for selecting the statistical significant lags. Hence, a series of speed of mean reversion parameters, $a_i(t)$, are estimated. By computing the derivative of the network output with respect to the network input, we obtain a series of daily values for $a_i(t)$. This is done for the first time, and it gives us a much better insight into DAWS dynamics and wind derivative pricing. As we will see, the daily variation of $a_i(t)$ is quite significant after all. In addition, it is expected that the waveform of the wavelet network will provide a better fit to the DATs that are governed by seasonalities and periodicities.

Using wavelet networks the generalized version of the autoregressive dynamics of detrended and deseasonalized DAWS estimated nonlinearly and nonparametrically is given by

$$\tilde{W}^{(l)}(t+1) = \phi\left(\tilde{W}^{(l)}(t), \tilde{W}^{(l)}(t-1), \ldots\right) + e(t) \tag{9.6}$$

where

$$e(t) = \sigma(t)\varepsilon(t) \tag{9.7}$$

and $\tilde{W}^{(l)}(t)$ is given by (9.3).

Variable Selection

Model (9.6) uses past DAWSs (detrended and deseasonalized) over one period. Using more lags, we expect to overcome the strong correlation found in the residuals. However, the length of the lag series must be selected. In previous chapters detailed explanations were given of how to apply the model identification framework. Model identification can be separated into two parts: model selection and variable significance testing. Since wavelet networks are nonlinear tools, criteria such as AIC or LLF cannot be used. Hence, in this section the framework proposed will be used to select the significant lags, to select the appropriate network structure, to train a wavelet network in order to learn the dynamics of the wind speeds, and finally, to forecast the future evolution of wind speeds.

The model identification algorithm uses a recurrent algorithm to simultaneously estimate the correct number of lags that must be used to model the wind speed dynamics and the architecture of the wavelet network.

Our backward selection algorithm examines the contribution of each available explanatory variable to the predictive power of the wavelet network. First, the prediction risk of the wavelet network is estimated as well as the statistical significance of each variable. If a variable is statistically insignificant, it is removed from the training set and the prediction risk and new statistical measures are estimated. The algorithm stops if all explanatory variables are significant. Hence, in each step of our algorithm, the variable with the larger p-value greater than 0.1 will be removed from the training set of our model. After each variable removal, a new architecture of the wavelet network will be selected and a new wavelet network will be trained. However, the correctness of the decision of removing a variable must be examined. This can be done by examining either the prediction risk or $\bar{R}^2$. If the new prediction risk is smaller than the new prediction risk multiplied by a threshold, the decision of removing the variable was correct. If the prediction risk increased more than the allowed threshold, the variable was reintroduced back to the model. We set this threshold at 5%. The statistical measure selected is the SBP. Our results indicate that the SBP fitness criterion was found to outperform alternative criteria significantly in the variable selection algorithm.

The variable selection framework proposed will be utilized for the transformed, detrended, and deseasonalized wind speeds in New York to select the length of the lag series. The target values of the wavelet network are the DAWSs. The explanatory variables are lagged versions of the target variable. The relevance of a variable to the model is quantified by the SBP criterion. Initially, the training set contains the dependent variable and seven lags. The analysis in the preceding section indicates that a training set with seven lags will provide all the necessary information of the ACF of the detrended and deseasonalized DAWSs. Hence, the initial training set consists of 7 inputs, 1 output, and 7293 training pairs.

Table 9.6 summarizes the results of the model identification algorithm for New York. Both the model selection and the variable selection algorithms are included in the table. The algorithm concluded in four steps and the final model contains only three variables (i.e., three lags). The prediction risk for the reduced model is 0.0937,

TABLE 9.6   Variable Selection with Backward Elimination in New York[a]

Step	Variable to Remove (Lag)	Variable to Enter (Lag)	Variables in Model	Hidden Units (Parameters)	n/p Ratio	Empirical Loss	Prediction Risk
—			7	1 (23)	317.4	0.0467	0.0938
1	7	—	6	1 (20)	365.0	0.0467	0.0940
2	5	—	5	1 (17)	429.4	0.0467	0.0932
3	6	—	4	2 (23)	317.4	0.0467	0.0938
4	4	—	3	2 (18)	405.6	0.0468	0.0937

[a]The algorithm concluded in four steps. In each step the following are presented: which variable is removed, the number of hidden units for the particular set of input variables and the parameters used in the wavelet network, the ratio between the parameters and the training patterns, the empirical loss, and the prediction risk.

while for the original model it was 0.0938, indicating that the predictive power of the wavelet network was slightly increased. On the other hand, the empirical loss increased slightly from 0.0467 for the initial model to 0.0468 for the reduced model, indicating that the explained variability (unadjusted) decreased slightly. Finally, the complexity of the network structure and the number of parameters were reduced significantly in the final model. The initial model needed 1 hidden unit and 7 inputs. Hence, 23 parameters were adjusted during the training phase, so the ratio of the number of training pairs n to the number of parameters p was 317.4. In the final model only 2 hidden units and 3 inputs were used, so only 18 parameters were adjusted during the training phase and the ratio of the number of training pairs n to the number of parameters p was 405.6.

The statistics for the wavelet network at each step are given in Table 9.7. The first part of the Table 9.7 reports the values of the SBP and its p-value; then various fitting criteria are reported. A closer inspection of the table reveals that the various error measures are reduced in the final model. However, the values of $\bar{R}^2$ are relatively small in all cases. This is due to the presence of large noise values compared to the small values of the underlying function.

In the final model, only three of the seven variables were used. The complexity of the model was reduced while the prediction power of the reduced model was increased. However, the in-sample obtained was a slightly poorer fit. The algorithm proposed suggests that a wavelet network needs only three lags to extract the autocorrelation from the data, whereas the linear model needed four lags. A closer inspection of Table 9.6 reveals that wavelet networks with three and four lags have the same predictive power in-sample and out-of-sample. Hence, we chose the simpler model.

TABLE 9.7   Step-by-Step Variable Selection in New York[a]

Variable	Full Model SBP	Full Model p-Value	Step 1 SBP	Step 1 p-Value	Step 2 SBP	Step 2 p-Value	Step 3 SBP	Step 3 p-Value	Step 4 SBP	Step 4 p-Value
7	0.0000	**0.8392**								
6	0.0000	0.7467	0.0000	0.4855	0.0000	**0.9167**				
5	0.0000	0.6799	0.0000	**0.9467**						
4	0.0000	0.5203	0.0000	0.7180	0.0000	0.2643	0.0000	**0.7480**		
3	0.0001	0.1470	0.0001	0.0000	0.0001	0.4706	0.0001	0.4719	0.0003	0.0000
2	0.0010	0.0469	0.0010	0.0000	0.0010	0.0000	0.0009	0.0000	0.0010	0.0168
1	0.0141	0.0000	0.0141	0.0000	0.0137	0.0000	0.0140	0.0000	0.0135	0.0000
MAE	0.2430		0.2430		0.2428		0.2430		0.2429	
MaxAE	1.7451		1.7453		1.7156		1.7541		1.6986	
NMSE	0.8832		0.8832		0.8833		0.8832		0.8834	
$\bar{R}^2$	11.68%		11.67%		11.67%		11.68%		11.65%	
Empirical loss	0.0467		0.0467		0.0467		0.0467		0.0468	
Prediction risk	0.0938		0.0940		0.0932		0.0938		0.0937	
Iterations	22		37		26		19		225	

[a]The SBP is the average for each variable of 50 bootstrapped samples, the standard deviation, and the p-value; SBP, sensitivity-based pruning; MAE, mean absolute error; MaxAE, maximum absolute error; NMSE, normalized mean squared error; MSE, mean square error; MAPE, mean absolute percentage error.

TABLE 9.8   Prediction Risk at Each Step of the Variable Selection Algorithm for the First 5 Hidden Units for New York

Step	Hidden Units				
	1	2	3	4	5
0	**0.09378**	0.09380	0.09379	0.09379	0.09380
1	**0.09403**	0.09404	0.09403	0.09406	0.09406
2	**0.09321**	0.09324	0.09325	0.09326	0.09327
3	0.09384	**0.09380**	0.09384	0.09387	0.09386
4	0.09370	**0.09367**	0.09368	0.09373	0.09379

Model Selection

In this section the appropriate number of hidden units is determined by applying the model selection algorithm. Table 9.8 shows the prediction risk for the first 5 hidden units at each step of the variable selection algorithm for the DAWSs in New York. It is clear that only 1 hidden unit is sufficient to model the detrended and deseasonalized DAWSs in New York at the first three steps. Similarly, 2 hidden units were needed for the last two steps.

Initialization and Training

After the training set and the correct topology of the wavelet network are selected, the wavelet network can be constructed and trained. In this case study the BE method is used to initialize the wavelet network. A wavelet basis is constructed by scanning the first four levels of the wavelet decomposition of the detrended and deseasonalized DAWSs in New York.

The wavelet basis consists of 205 wavelets. To reduce the number of wavelets in the wavelet basis, the wavelets that contain fewer than six sample points of the training data in their support are removed. The truncated basis contains 119 wavelet candidates. Applying the BE method, the wavelets are ranked in order of significance. Since only 2 hidden units are used in the architecture of the model, the best two wavelets are selected. The results of previous steps are similar. The MSE after the initialization was only 0.09420. Figure 9.9a presented the initialization of the final model using 2 hidden units. The initialization is very good, and the wavelet network converged after 225 iterations. The training stopped when the minimum velocity, 10^{-5}, of the training algorithm was reached. The fitting of the trained wavelet network is shown in Figure 9.9b.

Model Adequacy

In this section the model adequacy of the wavelet network is studied. The n/p ratio is 405.3, indicating that each parameter of the network corresponds to 405 values. Hence, we can safely conclude that overfitting was avoided.

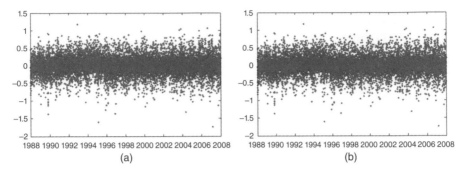

Figure 9.9 *Initialization of the final model for the wind data in New York using the BE method (a) and the fit of the trained network with 2 hidden units (b). The wavelet network converged after 225 iterations.*

TABLE 9.9   Residual Testing[a]

	Parameter	p-Values
n/p Ratio	405.3	–
Mean	0.0000	
Median	0.0018	
S. dev.	0.3058	
DW	2.0184	0.4526
LB Q-stat.	20.6002	0.4210
JB stat.	28.8917	0.0000
KS stat.	22.3453	0.0000

[a]S. dev., standard deviation; DW, Durbin–Watson; LB, Ljung–Box; KS, Kolmogorov–Smirnov.

In a closer examination of the residuals we found that the mean of the residuals is zero with a standard deviation of 0.3058. The normality hypothesis is rejected, but the hypothesis that the residuals are uncorrelated is accepted. The results are reported analytically in Table 9.9.

Various error criteria are reported in Table 9.10. The MSE is only 0.0937, while the NMSE and the RMSE are 0.8834 and 0.0937, respectively. Similarly, the maximum absolute error is 1.6986. Finally, the SMAPE is 27.55%. Note that since some values are zero, the MAPE cannot be computed. We observer that the SMAPE is relative high. This is due to the presence of a large error term compared to the very small values of the targets.

The parameters estimated for the regression between the target values and the network output are presented in Table 9.11. A closer inspection reveals that the

TABLE 9.10   Error Criteria[a]

Md.AE	MAE	MaxAE	SSE	RMSE	NMSE	MSE	MAPE	SMAPE
0.2034	0.2429	1.6986	682.4678	0.3058	0.8834	0.0937	–	27.55%

[a]Md.AE, median absolute error; MAE, mean absolute error; MaxAE, maximum absolute error; SSE, sum of squared errors; RMSE, root mean squared error; NMSE, normalized mean squared error; MSE, mean squared error; MAPE, mean absolute percentage error; SPAME, symmetric mean absolute percentage error.

TABLE 9.11 Regression Statistics[a]

	Parameter	p-Values	S.E.	T-Stat.
b_0	0.0001	0.9886	0.0036	0.0143
b_1	0.9655	0.0000	0.0311	31.0318
$b_1 = 1$ Test	0.9655	0.2670	0.0311	31.0318
F	962.9701	0.0000		
DW	1.9933	0.7611		

[a]S.E., squared error; DW, Durbin–Watson.

parameter b_0 is not statistically different from zero, while the parameter b_1 is not statistically different from 1 at significance level 5%. Moreover, the linear regression is statistically significant according to the F-statistic. Finally, as presented in Table 9.12, the changes in the direction metric are reported. More precisely, POCID, IPOCID, and POS are 69.64%, 43.28%, and 59.55%, respectively.

Speed of Mean Reversion and Seasonal Variance

The daily values of the speed of mean reversion function $a_i(t)$ (7293 values) are depicted in Figure 9.10. Since there are three significant lags, there are three mean-reverting functions. Our results indicate that the speed of mean reversion is not constant. On the contrary, its daily variation is quite significant; this fact naturally has an impact on the accuracy of the pricing equations and thus must be taken into account. Intuitively, it was expected that $a_i(t)$ would not be constant. If the wind speed today is away from the seasonal average, it is expected that the speed of mean reversion will be high (i.e., the wind speed cannot deviate from its seasonal mean for long periods).

Examining the second part of Figure 9.5, we conclude that the autocorrelation was removed from the data successfully; however, the seasonal autocorrelation in the squared residuals is still present, as shown in Figure 9.6. We remove the seasonal autocorrelation using equation (9.5). The estimated parameters are presented in Table 9.13 and, as expected, their values are similar to those of the case of the linear model. The empirical and fitted seasonal variance are presented in Figure 9.7. The variance is higher during the winter period and reaches its minimum during the summer period.

Finally, examining the final residuals of the wavelet network model, we observe that the distribution of the residuals is very close to the normal distribution shown in Figure 9.8 and the autocorrelation was removed from the data successfully. In addition, we observe an improvement in the distributional statistics, in contrast to the

TABLE 9.12 Change-in-Direction Metrics[a]

POCID	IPOCID	POS
69.64%	43.28%	59.55%

[a]POCID, prediction of change in direction; IPOCID, independent prediction of change in direction; POS, prediction of sign.

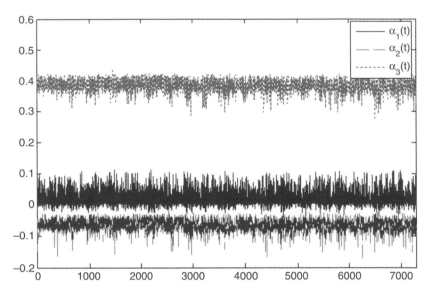

Figure 9.10 *Daily variations of the speed of mean reversion function $a_i(t)$ in New York.*

TABLE 9.13 Estimated Parameters of the Seasonal Variance for the WN

c_0	c_1	c_2	c_3	c_4	d_1	d_2	d_3	d_4
0.0935	−0.000020	−0.0034	0.0014	−0.0026	0.0353	−0.0016	−0.0042	−0.0052

case of the linear model. The distributional statistics of the residuals are presented in Table 9.14.

The distributional statistics of the residuals indicate that in-sample the two models can represent the dynamics of the DAWSs accurately; however, an improvement is evident when a nonlinear nonparametric wavelet network is used.

FORECASTING DAILY AVERAGE WIND SPEEDS

In this section the model proposed is validated out-of-sample. In addition, the performance of the model is tested against two models: first, against the linear model described previously, and second, against the simple persistent method, usually referred to as the benchmark. The linear model is the AR(4) model described in

TABLE 9.14 Descriptive Statistics of the Residuals for the Wavelet Network Model[a]

Var.	Mean	S. dev.	Max.	Min.	Skew	Kur.	JB	p-Value	KS	p-Value	LBQ	p-Value
Noise	0	1	3.32	−4.91	−0.08	3.04	8.84	0.0043	0.927	0.3544	13.437	0.858

S. dev., standard deviation; JB, Jarque–Bera statistic; KS, Komogorov–Smirnov statistic; LBQ, Ljung–Box Q-statistic.

the preceding section. The persistent method assumes that today's and tomorrow's DAWSs will be equal [i.e., $W^*(t + 1) = W(t)$, where W^* indicates the forecasted value].

The three models will be used for forecasting DAWSs for two different periods. Usually, wind derivatives are written for a period of a month. Hence, DAWSs for one and two months will be forecasted. The out-of-sample data set corresponds to the period from January 1 to February 28, 2008 and was not used for the estimation of the linear and nonlinear models. Note that our previous analysis reveals that the variance is higher in the winter period, indicating that it is more difficult to forecast DAWS accurately for these two months. To compare our results, the Monte Carlo approach is followed. We simulate 10,000 paths and calculate the average error criteria.

The performance of the three methods when the forecast window is one month is presented in Table 9.15. Various error criteria are estimated, such as the mean error, the median, Max. AE, the MSE, the POCID, and the IPOCID. As shown in the table, our proposed method outperforms both the persistent and the AR(4) model. The AR(4) model performs better than the naive persistent method; however, all error criteria are improved further when a nonlinear wavelet network is used. The MSE is 16.3848 for the persistent method, 10.5376 for the AR(4) model, and 10.4643 for the wavelet network. In addition, our model can predict the movements of the wind speed more accurately since the POCID is 80% for the wavelet network and the AR(4) models, whereas it is only 47% for the persistent method. Moreover, the IPOCID is 37% for the model proposed, whereas it is only 33% for the other two methods.

Next, the three forecasting methods are evaluated for two months of day-ahead forecasts. The results are similar and are presented in Table 9.16. The wavelet network proposed outperforms the other two methods. Only the Md.AE is slightly better when the AR(4) model is used. However, the IPOCID is 38% for the AR(4) method, whereas

TABLE 9.15 Out-of-Sample Comparison of One-Month Forecasts of DAWS[a]

	Persistent	AR(4)	WN
Md.AE	2.3000	2.3363	2.0468
ME	−0.0483	0.2117	−0.0485
MAE	3.3000	2.5403	2.5151
MaxAE	8.2000	7.9160	7.7019
SSE	507.9300	326.6666	324.3940
RMSE	4.0478	3.2461	3.2348
NMSE	1.5981	1.0278	1.0206
MSE	16.3848	10.5376	10.4643
MAPE	0.3456	0.2724	0.2680
SMAPE	0.3233	0.2555	0.2518
POCID	47%	80%	80%
IPOCID	33%	33%	37%
POS	100%	100%	100%

[a]Md.AE, median absolute error; ME, mean error; MAE, mean absolute error; MaxAE, maximum absolute error; SSE, sum of squared errors; RMSE, root mean squared error; NMSE, normalized mean squared error; MSE, mean squared error; MAPE, mean absolute percentage error; SMAPE, symmetric MAPE; POCID, position of change in direction; IPOCID, independent POCID; POS, position of sign.

TABLE 9.16   Out-of-Sample Comparison of Two-Month Forecasts of DAWS[a]

	Persistent	AR(4)	WN
Md.AE	2.4000	2.6393	2.6745
ME	0.1101	0.3570	0.1616
MAE	3.3678	2.8908	2.7967
MaxAE	11.2000	10.0054	8.3488
SSE	1054.3500	754.9589	688.2363
RMSE	4.2273	3.5771	3.4154
NMSE	1.4110	1.0103	0.9210
MSE	17.8703	12.7959	11.6650
MAPE	0.3611	0.3126	0.3056
SMAPE	0.3289	0.2808	0.2778
POCID	45%	69%	69%
IPOCID	36%	38%	43%
POS	100%	100%	100%

[a]Md.AE, median absolute error; ME, mean error; MAE, mean absolute error; MaxAE, maximum absolute error; SSE, sum of squared errors; RMSE, root mean squared error; NMSE, normalized mean squared error; MSE, mean squared error; MAPE, mean absolute percentage error; SMAPE, symmetric MAPE; POCID, position of change in direction; IPOCID, independent POCID; POS, position of sign.

it is 43% for the wavelet network. Also, our results indicate that the benchmark persistent method produces significantly poorer forecasts.

Our results indicate that the wavelet network can forecast evolution of the dynamics of the DAWSs and hence constitute an accurate tool for wind derivative pricing. The cumulative average wind speed (CAWS) index is calculated to provide better insight into the performance of each method. Since we are interested in weather derivatives, one common index is the sum of the daily average wind speed index over a specific period. An estimation of three methods is presented in Table 9.17. The wavelet network, the AR(4), and the historical burn analysis (HBA) method are compared. The HBA is a simple statistical method that estimates the performance of the index over a specific period in the past and is often used in the industry. It represents the average of 20 years of the index during January and February and serves as a benchmark.

The final row of Table 9.17 presents the actual values of the cumulative rainfall index. An inspection of the table reveals that the wavelet network significantly outperforms the other two methods. For the first case, where forecasts for one month ahead are estimated, the forecast of the CAWS index using the wavelet network is 312.7, while the actual index is 311.2. On the other hand, the forecast using the AR(4) model is 305.1. However, when the forecast period is increased, the forecast of the AR(4) model deviates significantly. For the second case, the forecast of the wavelet

TABLE 9.17   Estimation of the Cumulative Rainfall Index for 1 and 2 Months Using an AR(4) model, Wavelet Network, and Historical Burn Analysis

	AR(4)	WN	HBA	Actual
1 month	305.1	312.7	345.5	311.2
2 months	579.5	591.1	658.3	600.6

network is 591.1, the actual index is 600.6, and the AR(4) forecast is 579.5. Finally, we have to mention that the wavelet network uses less information than the AR(4) model, since with the wavelet network only the information of from three lags is used.

Since we are interested in wind derivatives and the valuation of wind contracts, an illustration of the performance of each method using a theoretical contract is presented next. A common wind contract has a tick size of 0.1 knot and pays \$20 per tick size. Hence, for a one-month contract the AR(4) method underestimates the contract size by \$1200, while the wavelet network only overestimates the contract by \$300. Similarly, for a two-month contract the AR(4) method underestimates the contract size by \$4220, while the wavelet network underestimates the contract by \$1900.

Incorporating meteorological forecasts can lead to a potentially significant improvement in the performance of the model proposed. Meteorological forecasts can easily be incorporated in both the linear and wavelet network models presented previously. A similar approach was followed for temperature derivatives by Dorfleitner and Wimmer (2010). However, this method cannot always be applied. Despite great advances in meteorological science, weather still cannot be predicted precisely and consistently, and forecasts beyond 10 days are not considered accurate (Wilks, 2011). If the day that the contract is traded is during or close to the life of the derivative (during the period that wind measurements are considered), the meteorological forecasts can be incorporated in order to improve the performance of the methods. However, very often, weather derivatives are traded long before the start of the life of the derivative. More precisely, very often weather derivatives are traded months or even a season before the starting day of the contract. In that case, meteorological forecasts cannot be used.

CONCLUSIONS

In this chapter the DAWSs from New York were studied. Our analysis revealed strong seasonality in the mean and variance. The DAWSs were modeled by a mean reverting Ornstein–Uhlenbeck process in the context of wind derivative pricing. In this study the dynamics of the wind-generating process are modeled using a nonparametric nonlinear wavelet network. Our proposed methodology was compared in-sample and out-of-sample against two methods often used in prior studies. The characteristics of the wind speed process are very similar to the process of daily average temperatures. Our results indicate a slight downward trend and seasonality in the mean and variance. In addition, the seasonal variance is higher in winter, reaching its lower values during the summer period.

Our method is validated in a two-month-ahead out-sample forecast period. Moreover, the various error criteria produced by the wavelet network are compared against the linear AR model and the persistent method. Results show that the wavelet network outperforms the other two methods, indicating that wavelet networks constitute an accurate model for forecasting DAWSs. More precisely, the wavelet network forecasting ability is stronger in both samples. When we test the fitted residuals of the wavelet network, we observe that the distribution of the residuals is very close to normal. Also, the wavelet network needed only the information of the past three days, while the linear method suggested a model with four lags. Finally, although we focused on DAWSs, our model can easily be adapted in hourly modeling.

The results in this case study are preliminary and can be analyzed further. More precisely, alternative methods for estimating the seasonality in the mean and in the variance can be developed. Alternative methods could improve the fitting to the original data as well as the training of the wavelet network. In addition, the inclusion of meteorological forecasts can further improve the forecasting performance of the wavelet networks.

It is also important to test the largest forecasting window of each method. Since meteorological forecasts of a window larger than a few days are considered inaccurate, this analysis will suggest the best model according to the forecasting interval desired.

Finally, a large-scale comparison must be conducted. Testing the methods proposed as well as more sophisticated models such as general ARFIMA or GARCH in various meteorological stations will provide a better insight into the dynamics of the DAWS as well as in the predictive ability of each method.

REFERENCES

Ailliot, P., Monbet, V., and Prevosto, M. (2006). "An autoregressive model with time-varying coefficients for wind fields." *Envirometrics*, 17, 107–117.

Alaton, P., Djehince, B., and Stillberg, D. (2002). "On modelling and pricing weather derivatives." *Applied Mathematical Finance*, 9, 1–20.

Alexiadis, M. C., Dokopoulos, P. S., Sahsamanoglou, H. S., and Manousaridis, I. M. (1998). "Short-term forecasting of wind speed and related electrical power." *Solar Energy*, 63(1), 61–68.

Barbounis, T. G., Theocharis, J. B., Alexiadis, M. C., and Dokopoulos, P. S. (2006). "Long-term wind speed and power forecasting using local recurrent neural network models." *IEEE Transactions on Energy Conversion,* 21(1), 273–284.

Benth, F. E., and Saltyte-Benth, J. (2009). "Dynamic pricing of wind futures." *Energy Economics*, 31, 16–24.

Beyer, H. G., Degner, T., Hausmann, J., Hoffmann, M., and Rujan, P. (1994). "Short-term prediction of wind speed and power outputof a wind turbine with neural networks." *2nd European Congress on Intelligent Techniques and Soft Computing*, Aachen, Germany.

Billinton, R., Chen, H., and Ghajar, R. (1996). "Time-series models for reliability evaluation of power systems including wind energy." *Microelectronics and Reliability*, 36(9), 1253–1261.

Brody, C. D., Syroka, J., and Zervos, M. (2002). "Dynamical pricing of weather derivatives." *Quantitave Finance*, 2, 189–198.

Brown, B. G., Katz, R. W., and Murphy, A. H. (1984). "Time-series models to simulate and forecast wind speed and wind power." *Journal of Climate and Applied Meteorology*, 23, 1184–1195.

Cao, M., and Wei, J. (2003). "Weather Derivatives: a New Class of Financial Instruments." Working Paper, University of Toronto, Toronto, Canada.

Caporin, M., and Pres, J. (2010). "Modelling and forecasting wind speed intensity for weather risk management." *Computational Statistics and Data Analysis.*

Castino, F., Festa, R., and Ratto, C. F. (1998). "Stochastic modelling of wind velocities time-series." *Journal of Wind Engineering and Industrial Aerodynamics*, 74–76, 141–151.

Celik, A. N. (2004). "A statistical analysis of wind power density based on the Weibull and Rayleigh models at the southern region of Turkey." *Renewable Energy*, 29(4), 593–604.

Cripps, E., Nott, D., Dunsmuir, W. T. M., and Wikle, C. (2005). "Space-time modelling of Sydney Harbour winds." *Australian and New Zealand Journal of Statistics*, 47(1), 3–17.

Daniel, A. R., and Chen, A. A. (1991). "Stochastic simulation and forecasting of hourly average wind speed sequences in Jamaica." *Solar Energy*, 46(1), 1–11.

Dorfleitner, G., and Wimmer, M. (2010). "The pricing of temperature futures at the Chicago Mercantile Exchange." *Journal of Banking and Finance.*

Dorvlo, A. S. S. (2002). "Estimating wind speed distribution." *Energy Conversion and Management*, 43(17), 2311–2318.

Garcia, A., Torres, J. L., Prieto, E., and de Francisco, A. (1998). "Fitting wind speed distributions: a case study." *Solar Energy*, 62(2), 139–144.

Haslett, J., and Raftery, A. E. (1989). "Space-time modelling with long-memory dependence: assessing Ireland's wind power resource." *Journal of the Royal Statistical Society,* Ser C, 38(1), 1–50.

Huang, Z., and Chalabi, Z. S. (1995). "Use of time-series analysis to model and forecast wind speed." *Journal of Wind Engineering and Industrial Aerodynamics*, 56, 311–322.

Jaramillo, O. A., and Borja, M. A. (2004). "Wind speed analysis in La Ventosa, Mexico: a bimodal probability distribution case." *Renewable Energy*, 29(10), 1613–1630.

Jewson, S., Brix, A., and Ziehmann, C. (2005). *Weather Derivative Valuation: The Meteorological, Statistical, Financial and Mathematical Foundations*, Cambridge University Press, Cambridge, UK.

Justus, C. G., Hargraves, W. R., Mikhail, A., and Graber, D. (1978). "Methods for estimating wind speed frequency distributions." *Journal of Applied Meteorology,* 17(3), 350–385.

Kamal, L., and Jafri, Y. Z. (1997). "Time-series models to simulate and forecast hourly averaged wind speed in Quetta, Pakistan." *Solar Energy*, 61(1), 23–32.

Kavak Akpinar, E., and Akpinar, S. (2005). "A statistical analysis of wind speed data used in installation of wind energy conversion systems." *Energy Conversion and Management*, 46(4), 515–532.

Kavasseri, R. G., and Seetharaman, K. (2009). "Day-ahead wind speed forecasting using f-ARIMA models." *Renewable Energy*, 34(5), 1388–1393.

Martin, M., Cremades, L. V., and Santabarbara, J. M. (1999). "Analysis and modelling of time-series of surface wind speed and direction." *International Journal of Climatology*, 19, 197–209.

Mohandes, M. A., Rehman, S., and Halawani, T. O. (1998). "A neural networks approach for wind speed prediction." *Renewable Energy*, 13(3), 345–354.

Mohandes, M. A., Halawani, T. O., Rehman, S., and Hussain, A. A. (2004). "Support vector machines for wind speed prediction." *Renewable Energy*, 29(6), 939–947.

More, A., and Deo, M. C. (2003). "Forecasting wind with neural networks." *Marine Structures*, 16, 35–49.

Nfaoui, H., Buret, J., and Sayigh, A. A. M. (1996). "Stochastic simulation of hourly average wind speed sequences in Tangiers (Morocco)." *Solar Energy*, 56(3), 301–314.

Nielsen, T. S., Madsen, H., Nielsen, H. A., Pinson, P., Kariniotakis, G., Siebert, N., Marti, I., Lange, M., Focken, U., Bremen, L. V., Louka, G., Kallos, G., and Galanis, G. (2006). "Short-term wind power forecasting using advanced statistical methods." *European Wind Energy Conference*, Athens, Greece.

Pinson, P., and Kariniotakis, G. N. (2003). "Wind power forecasting using fuzzy neural networks enhanced with on-line prediction risk assessment." *Power Tech Conference Proceedings IEEE Bologna*, Vol. 2.

Saltyte-Benth, J., and Benth, F. E. (2010). "Analysis and modelling of wind speed in New York." *Journal of Applied Statistics*, 37(6), 893–909.

Sfetsos, A. (2000). "A comparison of various forecasting techniques applied to mean hourly wind speed time-series." *Renewable Energy*, 21, 23–35.

Sfetsos, A. (2002). "A novel approach for the forecasting of mean hourly wind speed time series." *Renewable Energy*, 27, 163–174.

Tol, R. S. J. (1997). "Autoregressive conditional heteroscedasticity in daily wind speed measurements." *Theoretical and Applied Climatology*, 56, 113–122.

Tolman, H. L., and Booij, N. (1998). "Modeling wind waves using wavenumber–direction spectra and a variable wavenumber grid." *Global Atmosphere and Ocean System*, 6, 295–309.

Torres, J. L., Garcia, A., De Blas, M., and De Francisco, A. (2005). "Forecast of hourly average wind speed with ARMA models in Navarre (Spain)." *Solar Energy*, 79, 65–77.

Tuller, S. E., and Brett, A. C. (1984). "The characteristics of wind velocity that favor the fitting of a Weibull distribution in wind speed analysis." *Journal of Climate and Applied Meteorology*, 23(1), 124–134.

Wilks, D. S. (2011). *Statistical Methods in the Atmospheric Sciences*. Academic Press, Oxford, UK.

WRMA. (2010). "Weather derivatives volume plummets." Retrieved January, 2010, from www.wrma.org/pdf/weatherderivativesvolumeplummets.pdf

Yamada, Y. (2008). "Simultaneous optimization for wind derivatives based on prediction errors." *Americal Control Conference*, Washington, DC, 350–355.

Zapranis, A., and Alexandridis, A. (2008). "Modelling temperature time dependent speed of mean reversion in the context of weather derivetive pricing." *Applied Mathematical Finance*, 15(4), 355–386.

10

Predicting Chaotic Time Series

One of the most common problems that financial managers have to face is the accurate prediction of time series. Various models are used for the modeling of stock returns and prices or levels of volatility. The usual approach that financial analysts follow to solve these problems is to employ statistical procedures such as the general family of ARFIMA and/or GARCH models. However, financial time series often exhibit chaotic behavior. As a result, the prediction power of linear models is limited, due to their inability to model the evolutionary dynamics of the process. The models mentioned previously do not have the ability to adapt to a change of the dynamics in a chaotic system.

Chaotic time series are dynamic systems that are extremely sensitive to initial conditions and can exhibit complex external behavior. Small differences in initial conditions can exhibit diverging results. Hence, it is difficult to find the dynamic system simply through observations of the outcome. As defined by Edward Lorenz, in chaos the present determines the future, but the approximate present does not approximately determine the future.

To improve the modeling of chaotic time series, a number of nonlinear prediction methods have been developed, such as polynomials, neural networks, genetic algorithms, dynamic programming, and swarm optimization. Recently, neural networks and local models have been employed directly for chaotic time-series prediction, and comparatively satisfactory results have been reported (Inoue et al., 2001).

Wavelet Neural Networks: With Applications in Financial Engineering, Chaos, and Classification,
First Edition. Antonios K. Alexandridis and Achilleas D. Zapranis.
© 2014 John Wiley & Sons, Inc. Published 2014 by John Wiley & Sons, Inc.

In this chapter, wavelet networks are used to model the dynamics of a chaotic time series. More precisely, the framework proposed in earlier chapters is evaluated in a chaotic time series where the data were generated from the Mackey–Glass equation (Mackey and Glass, 1977). Moreover, the trained wavelet network is used to predict the future evolution of the chaotic system.

MACKEY–GLASS EQUATION

The Mackey–Glass equation was proposed by Mackey and Glass (1977) as a model of white blood cell production. Subsequently, it was popularized in the neural network field due to its richness in structure (Hsu and Tenorio, 1992). The equation was first proposed to associate the onset of disease with bifurcations in the dynamics of first-order differential-delay equations that model physiologic systems (Mackey and Glass, 1977).

The Mackey–Glass equation is a time-delay differential equation given by

$$\frac{\partial x}{\partial t} = \frac{ax(t - \tau)}{1 + x^c (t - \tau)} - bx(t) \tag{10.1}$$

The equation displays a broad diversity of dynamic behavior, including limit-cycle oscillation, with a variety of waveforms, and apparently aperiodic or "chaotic" solutions (Mackey and Glass, 1977). The behavior of the Mackey–Glass equation depends on the choice of the time delay τ. If $\tau < 4.53$, the behavior of the Mackey–Glass equation is characterized by a stable fixed-point attractor. Similarly, if $4.53 < \tau < 13.3$, there is a stable limit-cycle attractor. For $13.3 < \tau < 16.8$, the period of the limit cycle doubles. Finally, if $\tau > 16.8$, the behavior of the Mackey–Glass equation is characterized by a chaotic attractor, which is characterized by τ.

The common value used is $\tau = 17$, and this is also the case in this case study (Hsu and Tenorio, 1992; Yingwei et al., 1997). The Mackey–Glass equation with $\tau = 17$ has chaotic behavior and an attractor with a fractal dimension of about 2.1. The usual function approximation approaches are disadvantageous when the fractal dimension is greater than 2 (Cao et al., 1995; Iyengar et al., 2002). At $\tau = 17$ the time series appear to be quasiperiodic and the power spectrum is broadband with numerous spikes, due to the quasiperiodicity (Hsu and Tenorio, 1992).

Different values of the parameters a, b, and c can be used; however, the usual values of the parameters are $a = 0.2$, $b = 0.1$, and $c = 10$. The series is initialized at x_0 0.1. For the initialization transients to decay, the first 4000 data points were discarded (Platt, 1991; Yingwei et al., 1997). As in previous studies, the series is predicted with $v = 50$ sample steps ahead using four past samples: $x_{n-v}, x_{n-v-6}, x_{n-v-12}$, and x_{n-v-18}.

The chaotic Mackey–Glass differential delay equation is recognized as a benchmark problem that has been used and reported by a number of researchers for comparing the learning and generalization ability of different models (Chen et al., 2006).

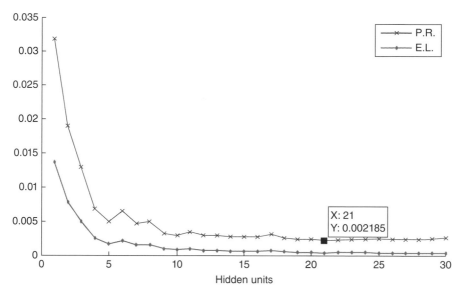

Figure 10.1 *Prediction risk (P.R.) and empirical loss (E.L.) for the first 30 hidden units for the Mackey–Glass equation.*

MODEL SELECTION

In this section the appropriate number of hidden units is determined by applying the model selection algorithm presented in Chapter 4. In Figure 10.1 the prediction risk and empirical loss for the Mackey–Glass equation are presented. The prediction risk was estimated up to a maximum of 35 hidden units. To estimate the prediction risk and find the optimal number of hidden units, the bootstrap method was used.

A close inspection of Figure 10.1 reveals that the empirical loss decreases (almost) monotonically as the complexity of the network increases. As expected, a better fit is obtained as more hidden units are used; however, the generalization ability does not necessarily increase. On the other hand, the prediction risk decreases (almost) monotonically until a minimum is reached and then will start to increase (almost) monotonically. The minimum value of the prediction risk, 0.002185, is obtained when a wavelet network with 21 hidden units is used. Hence, 21 hidden units were selected for the construction of the wavelet network model.

INITIALIZATION AND TRAINING

In the first two steps, the training set and correct topology of the wavelet network were selected. Next, the wavelet network can be constructed and trained. The training data set consists of 982 pairs. First, the wavelet network must be initialized. The

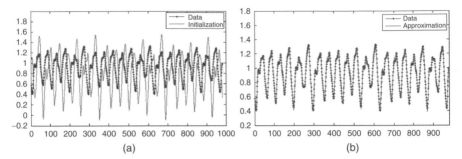

Figure 10.2 *(a) Initialization of the wavelet network using the backward elimination method and 21 hidden units; (b) approximation of the wavelet network after the training phase for the Mackey–Glass equation.*

backward elimination method was used for initialization of the set of parameters $w = (w_i^{[0]}, w_j^{[2]}, w_{\lambda+1}^{[2]}, w_{(\xi)ij}^{[1]}, w_{(\zeta)ij}^{[1]})$ of the wavelet network. Our results in Chapter 3 indicate that the BE method significantly outperforms alternative methods. A wavelet basis is constructed by scanning the first four levels of the wavelet decomposition of the data set.

The initial wavelet basis consists of 254 wavelets. However, not all wavelets in the wavelet basis contribute to the approximation of the original time series. The wavelets that contain fewer than six sample points of the training data in their support were removed. The truncated basis contains 116 wavelet candidates. The remaining wavelets were ranked and the "best" 21 wavelets were used for construction of the wavelet network.

First, we want to test if the initialization provides a good approximation of the Mackey–Glass function. To do so, the MSE between the initial approximation and the underlying function is computed. After the initialization the MSE was 0.220581, and the initialization needed 0.45 second to finish. The underlying function and the initialization of the wavelet network is shown in Figure 10.2a. A closer inspection of the figure reveals that the initialization is very good. The wavelet network converged after only 7277 iterations. The training stopped when the minimum velocity, 10^{-4}, of the training algorithm was reached. In this case the minimum velocity was increased slightly, from 10^{-5} to 10^{-4}, to avoid very large training times. The MSE error after training is 0.000299, and the total amount of time needed to train the network (initialization and training) was 39.2 seconds. $\bar{R}^2 = 99.30\%$, POCID = 89.60%, and IPOCID = 91.23%. The initialization of the WN and the final approximation after the training phase are presented in Figure 10.2.

MODEL ADEQUACY

Before proceeding to prediction of the time series out-of-sample, the model adequacy of the wavelet network will be studied. The n/p ratio is 5.06, indicating that each parameter of the network corresponds to five values. To avoid overfitting in problems

TABLE 10.1 Residual Testing[a]

	Parameter	p-Values
n/p Ratio	5.06	
Mean	0.0000	
Median	−0.0011	
S. dev.	0.0173	
DW	0.6238	0.0000
LB Q-stat.	954.3906	0.0000
JB stat.	25.7340	0.0000
KS stat.	15.0600	0.0000
R^2	99.44%	
$\bar{R}^2$	99.30%	

[a]S. dev., standard deviation; DW, Durbin–Watson; LB, Ljung–Box; KS, Kolmogorov–Smirnov.

where only a small number of observations are available, it is useful to look at the n/p ratio.

In a closer examination of the residuals we found that the mean of the residuals is zero with a standard deviation of 0.0173. The normality hypothesis is rejected as well as the hypothesis that the residuals are uncorrelated. This is not a problem for wavelet networks since, unlike linear models, normal distribution for the residuals was not assumed. Finally, the fitting of the wavelet network to the data is very good, with $R = 99.44\%$ and $\bar{R} = 99.30\%$. The results are reported analytically in Table 10.1.

The various error criteria are reported in Table 10.2. A close inspection of the table confirms our previous results that the fit is very good. The MSE is only 0.0003, while the NMSE and RMSE are only 0.0056 and 0.0173, respectively. Similarly, the maximum absolute error is 0.0673. Finally, the MAPE and SMAPE are 1.60% and 0.80%, respectively.

The estimated parameters of the regression between the target values and the network output are presented in Table 10.3. A close inspection reveals that the parameter b_0 is not statistically different from zero while the parameter b_1 is not statistically different from 1 at significance level 0.05. Moreover, the linear regression is statistically significant according to the F-statistic. Figure 10.3 is a scatter plot between the target values and the network output. Finally, as evident from Table 10.4, the change in direction metrics is very high. More precisely, POCID, IPOCID, and POS are 89.60%, 91.23%, and 100%, respectively.

TABLE 10.2 Error Criteria[a]

Md.AE	MAE	MaxAE	SSE	RMSE	NMSE	MSE	MAPE	SMAPE
0.0111	0.0134	0.0673	0.2935	0.0173	0.0056	0.0003	1.60%	0.80%

[a] Md.AE, median absolute error; MAE, mean absolute error; MaxAE, maximum absolute error; SSE, sum of squared errors; RMSE, root mean squared error; NMSE, normalized mean squared error; MSE, mean squared error; MAPE, mean absolute percentage error; SPAME, symmetric mean absolute percentage error.

TABLE 10.3   Regression Statistics[a]

	Parameter	p-Values	S.E.	T-Stat.
b_0	−0.0040	0.0774	0.0023	−1.7676
b_1	1.0044	0.0000	0.0024	417.6064
$b_1 = 1$ Test	1.0044	0.0687	0.0024	1.8223
R^2	99.44%			
F	174,391.7457	0.0000		
DW	0.6310	0.0000		

[a]S.E., squared error; DW, Durbin–Watson.

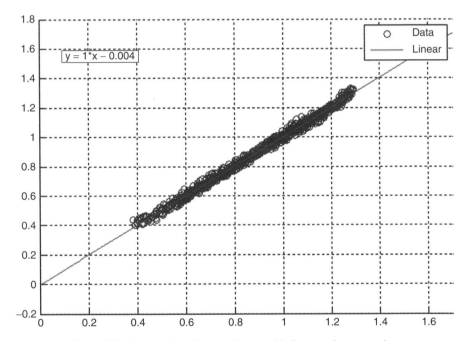

Figure 10.3   *Scatter plot of the wavelet network's fit versus the target values.*

TABLE 10.4   Change-in-Direction Metrics[a]

POCID	IPOCID	POS
89.60%	91.23%	100%

[a]POCID, prediction of change in direction; IPOCID, independent prediction of change in direction; POS, prediction of sign.

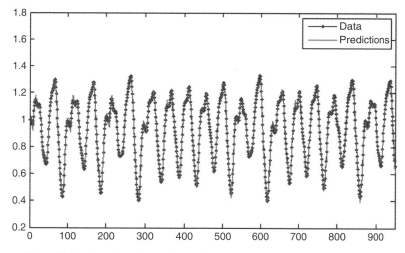

Figure 10.4   *Out-of-sample prediction of the wavelet network using 21 hidden units.*

PREDICTING THE EVOLUTION OF THE CHAOTIC MACKEY–GLASS TIME SERIES

In this section the performance of the wavelet network out-of-sample is evaluated. The training data set consists of 982 pairs and the out-of-sample data set consists of 950 pairs. These additional data were not used in training the wavelet network. The ability of the wavelet network to forecast the evolution of the chaotic Mackey–Glass equation is presented in Figure 10.4. Examining the figure, we can conclude that the wavelet network has very good generalization and forecasting ability, since the values predicted for the wavelet network are very close to the real target values of the Mackey–Glass equation.

Various statistics for the out-of-sample residuals are presented in Table 10.5. The residuals exhibit the same dynamics as in the training sample. The mean is close to zero and the standard deviation is 0.0186. The normality hypothesis is rejected as

TABLE 10.5   Out-of-Sample Residual Testing[a]

	Parameter	p-Values
n/p Ratio	4.89	
Mean	0.0001	
Median	–0.0002	
S. dev.	0.0186	
DW	0.5755	0.0000
LB Q stat.	750.1360	0.0000
JB stat.	46.4359	0.0000
KS stat.	14.7923	0.0000
R^2	99.32%	
$\bar{R}^2$	99.15%	

[a]S. dev., standard deviation; DW, Durbin–Watson; LB, Ljung–Box; KS, Kolmogorov–Smirnov.

TABLE 10.6   Out-of-Sample Error Criteria.[a]

Md.AE	MAE	MaxAE	SSE	RMSE	NMSE	MSE	MAPE	SMAPE
0.0110	0.0142	0.0699	0.3294	0.0186	0.0068	0.0003	1.65%	0.83%

[a]Md.AE, median absolute error; MAE, mean absolute error; MaxAE, maximum absolute error; SSE, sum of squared errors; RMSE, root mean squared error; NMSE, normalized mean squared error; MSE, mean squared error; MAPE, mean absolute percentage error; SPAME, symmetric mean absolute percentage error.

TABLE 10.7   Out-of-Sample Regression Statistics[a]

	Parameter	p-Values	S.E.	T-Stat.
b_0	−0.0061	0.0177	0.0026	−2.3767
b_1	1.0067	0.0000	0.0027	374.0211
$b_1 = 1$ Test	1.0067	0.0135	0.0027	2.4751
R^2	99.33%			
F	139,891.7574	0.0000		
DW	0.5867	0.0000		

[a]S.E., squared error; DW, Durbin–Watson.

well as the hypothesis that the residuals are uncorrelated. Finally, the accuracy of the predictions of the wavelet network is very good with $R = 99.32\%$ and $\bar{R} = 99.15\%$. The results are reported analytically in Table 10.5.

Furthermore, the very good predictive power of the wavelet network is verified by the various error criteria in Table 10.6. The MSE in the out-of-sample data set is only 0.000347, and the NMSE and RMSE are only 0.00686, and 0.0186, respectively. Similarly, the maximum absolute error is 0.0699. Finally, MAPE and SMAPE are 1.65% and 0.83%, respectively. Additional error criteria are reported in Table 10.6.

Our results so far indicate that the wavelet network proposed can accurately forecast the evolution of a chaotic time series such as the Mackey–Glass equation.

The estimated parameters of the regression between the target values and the network output are presented in Table 10.7. The scatter plot between the out-of-sample target values and the forecasted values of the wavelet network is presented in Figure 10.5. In both cases it is clear that the values forecast and the target values are similar. Moreover, the linear regression is statistically significant according to the F-statistic. Finally, as shown in Table 10.8, the change-in-direction metrics are very high. More precisely, POCID, IPOCID, and POS are 87.36%, 90.41%, and 100%, respectively, indicating that the wavelet network can predict with great accuracy the changes in the direction of the chaotic dynamic system.

CONFIDENCE AND PREDICTION INTERVALS

After the wavelet network is constructed and trained, it can be used for prediction. However, in many applications, and especially in finance, risk managers may be more interested in predicting intervals for future movements of the underlying function than

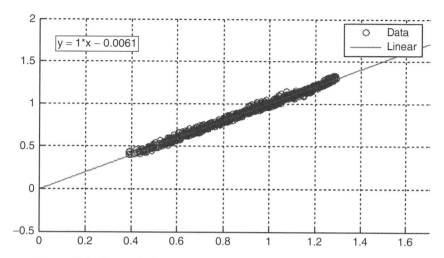

Figure 10.5   *Scatter plot between the wavelet network's output and the target values.*

simply point estimates. Hence, confidence and prediction intervals can be constructed. In this section both confidence and prediction intervals are constructed using the balancing method. Using the BS method, 200 training samples are created and divided into eight groups. In each group the average output of the wavelet networks is estimated. Next, 1000 new bootstrapped samples are created for the eight average outputs to estimate the model variance given by (7.25). Then the confidence intervals are estimated with a level of significance a of 5%. Unlike the previous case studies and examples, in this case no noise is added to the underlying function. Moreover, the network fitting and prediction are very good, as mentioned in the preceding section, with $\bar{R}^2 = 99.30\%$ in-sample and $\bar{R}^2 = 99.15\%$ out-of-sample. As a result, it is expected that the variance $\sigma_p^2 = \sigma_m^2 + \sigma_\varepsilon^2$ will be very small.

Figure 10.6 presents the confidence intervals and the true underlying function, which is the Mackey–Glass equation. Since the confidence intervals are very narrow, for clarity only one section is shown in Figure 10.6. It is clear that the underlying function is always between the confidence intervals.

In addition, in Figure 10.7 the prediction intervals for the out-of-sample data set together with the data read and the average forecast of the wavelet network for the 200 bootstrapped samples are presented. PICP = 98.8%.

TABLE 10.8   Out-of-Sample Change in Direction Metrics[a]

POCID	IPOCID	POS
87.36%	90.41%	100.00%

[a] POCID, prediction of change in direction; IPOCID, independent prediction of change in direction; POS, prediction of sign.

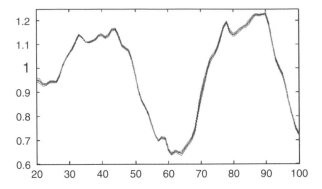

Figure 10.6 *In-sample confidence intervals (light gray line) together with the Mackey–Glass equation (black line) using the balancing method.*

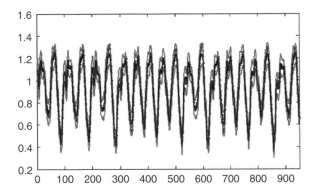

Figure 10.7 *Out-of-sample prediction intervals (light gray lines) together with the Mackey–Glass equation (gray line) and the network approximation (black line) using the balancing method (PICP = 98.8%).*

CONCLUSIONS

In this chapter the ability of a wavelet network to learn and predict the dynamics of a chaotic system was tested using the Mackey–Glass equation. The chaotic Mackey–Glass differential delay equation is recognized as a benchmark problem that has been used and reported by a number of researchers for comparing the learning and generalization ability of different models.

The objective was to predict $v = 50$ steps ahead using four past samples: $x_{n-v}, x_{n-v-6}, x_{n-v-12},$ and x_{n-v-18}. Following the framework proposed in this book, a wavelet network with 21 hidden units was constructed. The optimal topology of the network was selected by estimating the minimum prediction risk criterion using the bootstrap method. The wavelet network model was initialized by applying the backward elimination method.

The trained wavelet network provided a very good fit to the data. Moreover, the performance of the wavelet network in the out-of-sample data was very good. The accuracy of the predictions of the wavelet network was very good, with $R = 99.32\%$ and $\bar{R} = 99.15\%$.

Our results indicate that the wavelet network can predict the future evolution of the chaotic system accurately as well as predict the change in the direction of the chaotic Mackey–Glass dynamic system.

REFERENCES

Cao, L., Hong, Y., Fang, H., and He, G. (1995). "Predicting chaotic time series with wavelet networks." *Physica, Ser D*, 85, 225–238.

Chen, Y., Yang, B., and Dong, J. (2006). "Time series prediction using a local linear wavelet neural wavelet." *Neurocomputing*, 69, 449–465.

Hsu, W., and Tenorio, M. F. (1992). "Plastic network for predicting the Mackey–Glass time series." *International Joint Conference on Neural Networks*, Baltimore, MD.

Inoue, H., Fukunaga, Y., and Narihisa, H. (2001). "Efficient hybrid neural network for chaotic time series prediction." In *Artificial Neural Networks—ICANN* Springer-Verlag, New York, 712–718.

Iyengar, S. S., Cho, E. C., and Phoha, V. V. (2002). *Foundations of Wavelet Networks and Applications*. CRC Press, Grand Rapids, MI.

Mackey, M. C., and Glass, L. (1977). "Oscillation and chaos in physiological control systems." *Science*, 197(4300), 287–289.

Platt, J. (1991). "A resource-allocating network for function interpolation." *Neural Computation*, 3, 213–225.

Yingwei, L., Sundararajan, N., and Saratchandran, P. (1997). "A sequential learning scheme for function approximation using minimal radial basis function neural networks." *Neural Computation*, 9(2), 461–478.

11

Classification of Breast Cancer Cases

Breast cancer has become a major cause of death among women in developed countries (Senapati et al., 2011). As the causes of breast cancer remain unknown, early detection is crucial to reduce the death rate. However, early detection requires accurate and reliable diagnosis (Cheng et al., 2010). A diagnostic tool should distinguish between benign and malignant tumors while producing low false-positive (rate of missing chances) and false-negative (rate of failure) rates.

Mammography is probably the most effective method for breast tumor detection. However, the technique has limitations in cancer detection. For example, due to its low specificity, many unnecessary biopsy operations are performed, increasing the cost, the emotional pressure, and in some cases the risk to the patient (Zainuddin and Ong, 2010).

In this chapter a wavelet network is constructed to classify breast cancer based on various attributes. Hence, a computer-aided system is developed and proposed to provide additional accuracy in the classification of benign and malignant cases of breast tumors. The Wisconsin breast cancer (WBC) data set was obtained by the UCI Machine Learning Repository and was provided by Mangasarian and Wolberg (1990). In this particular case study we are more interested in producing fewer false negatives. Whereas a false positive will result in extra cost for additional clinical tests, a false negative may result in the death of the patient.

Wavelet Neural Networks: With Applications in Financial Engineering, Chaos, and Classification,
First Edition. Antonios K. Alexandridis and Achilleas D. Zapranis.
© 2014 John Wiley & Sons, Inc. Published 2014 by John Wiley & Sons, Inc.

Cheng et al. (2010) detected breast cancer based on ultrasound images. They employed a variety of classifiers, such as wavelet networks, and neural networks, and support vector machines to construct a computer-aided diagnostic system.

Similarly, Zainuddin and Ong (2010) used microarray slides as inputs to a wavelet neural network for cancer diagnosis. El-Sebakhy et al. (2006) proposed and evaluated functional networks from the WBC data set. Their results indicate that their proposed classifier is reliable and efficient.

Senapati et al. (2011) proposed a local linear wavelet neural network for breast cancer recognition. Their model was evaluated on the WBC data set and compared with methods already developed. Their results indicate that the local linear wavelet neural network performs better and has a higher level of generalization than those of common existing approaches. Methods other than artificial intelligence have been proposed for breast cancer classification: for example, linear programing by Mangasarian and Wolberg (1990) and fuzzy logic by Hassanien and Ali (2006).

In the remainder of the chapter we evaluate the classification ability of a wavelet network using two methods. In the first, the wavelet network is trained on the training sample and then evaluated out-of-sample in the validation sample. In the second, a cross-validation technique is utilized for training and forecasting evaluation of the wavelet network. Moreover, the model identification method is used to find the optimal set of input variables and the optimal structure of the wavelet network.

Data

The Wisconsin breast cancer (WBC) data set contains 699 samples. However, 16 values are missing, reducing the sample to 683 values. Each instance has one of two possible classes: benign or malignant. There are 239 (35%) malignant cases and 444 (65%) benign cases. The aim is to construct a wavelet network that classifies each clinical case accurately. The classification is based on nine attributes: clump thickness, uniformity of cell size, uniformity of cell shape, marginal adhesion, single epithelial cell size, bare nuclei, bland chromatin, normal nucleoli, and mitoses.

PART A: CLASSIFICATION OF BREAST CANCER

In this case the data set was split into training and validation samples. The training sample consists of 478 (70%) cases. The validation sample, consisting of 205 (30%) cases, is used to evaluate the predictive and classification power of the trained wavelet network. We assume that all nine variables are statistically significant, and they will be used as predictors. Hence, the variable selection algorithm is omitted in this section.

Model Selection

To construct the wavelet network, first the optimal number of hidden units must be found. To do so, the minimum prediction risk criterion is applied and the bootstrap method will be used. From the training sample we created 50 bootstrapped samples.

The prediction risk was minimized when only 1 hidden unit was used. The prediction risk for the full model was 0.2796 and the empirical loss was 0.0862.

Initialization and Training

The BE method is used to initialize the wavelet network. A wavelet basis is constructed by scanning the first four levels of the wavelet decomposition of the data. The wavelet basis is very large and consists of 719 wavelets. However, not all wavelets in the wavelet basis contribute to the approximation of the original time series. The wavelets that contain fewer than 11 sample points of the training data in their support are removed. Seven hundred and seven wavelets that do not contribute significantly to the approximation of the original time series were identified. The truncated basis contains only 12 wavelet candidates. Applying the BE method, the wavelets are ranked in order of significance. Since only 1 hidden unit is used on the architecture of the model, only the wavelet with the highest ranking is used to initialize the wavelet network. The initialization was very good and the MSE after the initialization was only 0.1905. The time needed for the initialization was 0.1692 second. The training stopped after 135 iterations, when the minimum velocity was reached and the MSE was 0.1725. The complete training time (initialization and training) was only 0.7572 second.

Classification

In-Sample   In the training sample there are 284 benign cases given the value -1 and 194 malignant cases given the value 1. The cutting score is -0.189. The classification matrix of the training sample is presented in Table 11.1. A closer inspection of the table reveals very good classification rates. More specifically, the wavelet network classified correctly 274 benign cases and 194 malignant cases. Hence, the wavelet network classified correctly 461 of 478 cases (96.44%). The specificity of the models is 96.39% and the sensitivity is 96.48%. On the other hand, the rate of failure and the rate of missing chances are very low, 3.61% and 3.52%, respectively. Finally, the fitness function is 0.5643.

Evaluation of the classification ability of the wavelet network is presented in Table 11.2. The maximum chance criterion is 59.41%, while in the heuristic method presented by Hair et al. (2010) it is 74.27%. Finally, the proportional chance criterion is 51.77%. The hit ratio is 96.44%, significantly larger than the various chance criteria.

TABLE 11.1   In-Sample Classification Matrix

Target	Forecast				
	Benign	Malignant	Total	Sensitivity	Specificity
Benign	274	10	284	96.48%	96.39%
Malignant	7	187	194	Rate of Missing Chances	Rate of Failure
Total	281	197	478	3.52%	3.61%

TABLE 11.2   Evaluation of the Classification Ability of the Wavelet Network

Maximum Chance	1.25% Max. Chance	Pro	Press's Q^a	Hit Ratio
59.41%	74.27%	51.77%	412.41	96.44%

^{a}Press's Q critical values at the confidence levels 0.1, 0.05, and 0.01: 2.71, 3.84, 6.63.

TABLE 11.3   Out-of-Sample Classification Matrix

	Forecast				
Target	Benign	Malignant	Total	Sensitivity	Specificity
Benign	159	1	160	99.38%	100%
Malignant	0	45	45	Rate of Missing Chances	Rate of Failure
Total	159	46	205	0.63%	0.00%

Hence, the model is predicting significantly better than chance. This is also confirmed by the large value of Press's Q statistic, which is greater than the critical values in the 0.1, 0.05, and 0.01 confidence levels.

Out-of-Sample   Next, the forecasting and classification ability of the trained wavelet network is evaluated in the validation sample. The data in the validation sample were not used during the training phase. Hence, these are new data that were never presented to the wavelet network. In the validation sample there are 160 benign cases given the value −1, and 45 malignant cases given the value 1. The classification matrix of the training sample is presented in Table 11.3. A close inspection of the table reveals the very good predictive ability and classification rates. More specifically, the wavelet network classified correctly 159 benign cases and 45 malignant cases. Hence, the wavelet network classified correctly 204 of 205 cases (99.51%). The specificity of the model is 100% and the sensitivity is 99.38%. Note that in this application the rate of failure is significantly more important than the rate of missing chances. If the network classifies a benign case as malignant, it is just a false alarm; however, classifying a malignant case as benign is lethal. Our results indicate that the wavelet network has very strong classification ability since the rate of failure and the rate of missing chances are very low: 0% and 0.63%, respectively. Finally, the fitness function is 0.5964.

Evaluation of the classification ability of the wavelet network is presented in Table 11.4. The maximum chance criterion is 78.05%, while in the heuristic method presented by Hair et al. (2010) it is 97.56%. Finally, the proportional chance criterion

TABLE 11.4   Out-of-Sample Evaluation of the Classification Ability of the Wavelet Network

Maximum Chance	1.25% Max. Chance	Pro	Press's Q^a	Hit Ratio
78.05%	97.56%	65.73%	201.02	99.51%

^{a}Press's Q critical values at the confidence levels 0.1, 0.05, and 0.01: 2.71, 3.84, 6.63.

is 65.73%. The hit ratio, 99.51%, is significantly larger than the maximum chance and the proportional chance criteria. Hence, the model is predicting significantly better than chance. This is also confirmed by the large value of Press's Q-statistic, which is greater than the critical values at the 0.1, 0.05, and 0.01 confidence levels.

PART B: CROSS-VALIDATION IN BREAST CANCER CLASSIFICATION IN WISCONSIN

In this section a different approach is followed. Instead of splitting the data into training and validation samples, cross-validation methods are used. In addition, we employ the variable selection algorithm to test if some of the attributes can be removed. More precisely, the model identification algorithm is utilized, and at each step the significant variables and the optimal structure of the wavelet network are estimated.

As mentioned earlier, cross-validation is used to assess the predictive power of the wavelet network. Hold-one-out cross-validation is used in each step of our algorithm. One training pattern of the data set will be removed from the training sample in each step. Then a wavelet network is trained on the remaining data. Finally, the trained network is evaluated on the pattern that was removed from the sample. The procedure is repeated 683 times, once from each training pattern.

Variable Selection

The target values of the wavelet network are the two possible classes. The explanatory variables are the nine attributes named earlier. To construct an accurate wavelet network classifier, the contribution of each attribute to the predictive power of the classifier must be tested. First, the significance of each attribute is examined. Hence, the initial training set consists of 9 inputs, 1 output, and 683 training samples. Again, the relevance of each attribute is quantified by the SBP criterion. Applying the variable selection proposed, the final model has only six variables, while the predictive power of the model remains almost unchanged.

Table 11.5 summarizes the results of the model identification algorithm for the WBC data. Both the model selection and variable selection algorithm are included

TABLE 11.5 Variable Selection with Backward Elimination in Wisconsin Breast Cancer Data Set[a]

Step	Variable to Remove	Variable to Enter (Lag)	Variables in Model	Hidden Units (Parameters)	n/p Ratio	Empirical Loss	Prediction Risk
—	—	—	9	1 (29)	23.6	0.0713	0.1488
1	X_9	—	8	1 (26)	26.7	0.0713	0.1485
2	X_4	—	7	3 (53)	12.9	0.0404	0.1136
3	X_3	—	6	3 (46)	14.8	0.0426	0.1135

[a]The algorithm concluded in four steps. In each step the following are presented: which variable is removed, the number of hidden units for the particular set of input variables and the parameters used in the wavelet network, the empirical loss, and the prediction risk.

TABLE 11.6   Step-by-Step Variable Selection[a]

	Full Model		Step 1		Step 2		Step 3	
Variable	SBP	p-Value	SBP	p-Value	SBP	p-Value	SBP	p-Value
1	0.0323	0.0000	0.0328	0.0000	0.0369	0.2695	0.0511	0.0000
2	0.0188	0.0000	0.0198	0.0000	0.0255	0.5068	0.0501	0.0000
3	0.0099	0.0873	0.0091	0.0000	0.0067	**0.6547**		
4	0.0025	0.0000	0.0023	**0.2166**				
5	0.0021	0.0985	0.0021	0.1995	0.0100	0.1655	0.0359	0.0000
6	0.1087	0.0000	0.1101	0.0000	0.1636	0.0000	0.01915	0.0000
7	0.0084	0.0831	0.0084	0.0945	0.0050	0.6080	0.0200	0.0000
8	0.0126	0.0000	0.0123	0.0298	0.0318	0.0000	0.0828	0.0000
9	0.0001	**0.9133**						
MAE	0.2474		0.2481		0.1473		0.2510	
MaxAE	1.6811		1.6481		1.9735		1.6857	
NMSE	0.1566		0.1569		0.0935		0.1597	
MAPE	24.73%		24.81%		14.73%		25.10%	
$\bar{R}^2$	83.67%		83.71%		89.88%		82.90%	
Empirical loss	0.0713		0.0713		0.0404		0.0426	
Prediction risk	0.1488		0.1485		0.1136		0.1135	
Iterations	149		119		11,283		264	

[a]SBP is the average SBP for each variable of 50 bootstrapped samples, the standard deviation, and the p-value. SBP, sensitivity-based pruning; MAE, mean absolute error; MaxAE, maximum absolute error; NMSE, normalized mean squared error; MSE, mean squared error; MAPE, mean absolute percentage error.

in the table. The algorithm was concluded in three steps and the final model consists of six variables only. A closer inspection of the table reveals that the empirical loss decreased from 0.0713 in the full model to 0.0426 in the reduced and simpler model. In addition, the prediction risk decreased from 0.1488 to 0.1135, indicating that the reduced model provides a better fitting to the data but also has better forecasting ability. The results of the variable significance algorithm indicate that the uniformity of cell shape, marginal adhesion, and mitoses should be removed from the input of the training sample in breast tumor classification. On the other hand, the attributes clump thickness, uniformity of cell size, single epithelial cell size, bare nuclei, bland chromatin, and normal nucleoli are statistically significant predictors.

Finally, the reduced model needed more hidden units in order to train the wavelet network. More precisely, in the full model only 1 hidden unit was used, corresponding to a 23.6 n/p ratio, while in the reduced model 3 hidden units were used, corresponding to a ratio of 14.8.

The statistics for the wavelet network model at each step are given in Table 11.6. The first part of the table reports the value of the SBP and its p-value. Various fitting criteria are also reported: MAE, MaxAE, NMSE, MAPE, $\bar{R}^2$, empirical loss, and prediction risk.

In the full model it is clear that the value of the SBP for the last variable (mitoses) is very high compared to the remaining variables. Observing the p-values, we conclude that the p-value of mitoses is 0.9133 and is greater than 0.1, strongly indicating a "not significant" variable. The wavelet network was converged after 149 iterations. In general, a very good fit was obtained. The empirical loss is 0.0713 and the prediction risk is 0.1488. MaxAE is 1.6811, MAE is 0.2474, and NMSE is 0.1566. MAPE is 24.74%. Finally, $\bar{R}^2 = 84.34\%$.

The statistics for the wavelet network at step 1 are also presented in Table 11.6. The network had 8 inputs, one wavelet was used to construct the wavelet network, and 26 weights were adjusted during the training phase. The wavelet network converged after 119 iterations. By removing X_9 from the model, we observe from Table 11.6 that the p-value of X_5 and X_4 became 0.1995 and 0.2166, respectively. The empirical loss remained the same. However, MAE and NMSE were increased slightly to 0.2481 and 0.1569, respectively. Similarly, the remaining error criteria were increased. Next, the decision to remove X_9 is tested. The new prediction risk was reduced to 0.1485, while the explained variability adjusted for degrees of freedom increased to 83.71%. Hence, the removal of X_6 reduced the complexity of the model while its predictive power increased.

At step 2, X_4(marginal adhesion), which had the largest p-value, 0.2166, at step 1, was removed from the model. The new wavelet network had 7 inputs, 3 hidden units were used for the architecture of the wavelet network, and 53 weights were adjusted during the training phase. The wavelet network converged after 11,283 iterations. All error criteria were reduced significantly and the new $\bar{R}^2$ is 89.88%. The new prediction risk is reduced significantly, to 0.1136. Hence, removing X_4, we obtain a better fit and better potential forecasting ability. Finally, observing the p-values, we conclude that at step 2, X_3 has the higher p-value, 0.6547.

In the final step the variable X_3 (uniformity of cell shape) was removed from the model. The network had 6 inputs, three wavelets were used for the construction of the wavelet network, and 46 weights were adjusted during the training phase. The wavelet network converged after 264 iterations. The new empirical loss was increased slightly to 0.0424, compared to 0.0404 in the previous step. Similarly, all error criteria were increased slightly. However, the explained variability adjusted for degrees of freedom was reduced to 82.90%. Hence, removing X_3, a slightly poorer fit was obtained. On the other hand, the prediction risk decreased further, to 0.1135.

The p-values of the remaining variables are zero, indicating that the remaining variables are characterized as very significant variables. Hence, the algorithm stops.

Model Selection

In each step of the algorithm the optimal number of hidden units is determined by applying the model selection algorithm. The results of the model selection algorithm are presented in Table 11.7. In the full model a wavelet network with 1 hidden unit was constructed. Applying the model selection algorithm using 50 bootstrapped samples of the initial training set, the prediction risk was minimized when only 1 hidden unit was used. The prediction risk for the full model was 0.1488 and the empirical loss was 0.0713. In the second step, the prediction risk increases monotonically again as the complexity of the wavelet network increases. Hence, the prediction risk is minimized when only 1 hidden unit is used and is 0.1485. Similarly, in the third step the prediction risk is minimized when 3 hidden units are used. In the final step,

TABLE 11.7   Prediction Risk at Each Step of the Variable Selection Algorithm for the First 5 Hidden Units

Step	Hidden Units				
	1	2	3	4	5
0	**0.1488**	0.1495	0.1504	0.1520	0.1519
1	**0.1485**	0.1550	0.1624	0.1699	0.1706
2	0.1526	0.1672	**0.1136**	0.1146	0.1232
3	0.1490	0.1573	**0.1135**	0.1157	0.1151

the reduced model needed 3 hidden units and the prediction risk was 0.1135. The empirical loss was 0.0426, indicating that the reduced model provides a better fit to the data but also has better forecasting ability.

Initialization and Training

After the training set and the correct topology of the wavelet network are selected, the wavelet network can be constructed and trained. The BE method is used to initialize the wavelet network. A wavelet basis is constructed by scanning the first four levels of the wavelet decomposition of the data set.

The initial wavelet basis consists of 675 wavelets. However, not all wavelets in the wavelet basis contribute to the approximation of the original time series. The wavelets that contain fewer than eight sample points of the training data in their support are removed. The truncated basis contains 28 wavelet candidates. The MSE after the initialization was 0.173170 and the initialization needed 0.23 second to finish. The initialization is very good and the wavelet network converged after only 264 iterations. The training stopped when the minimum velocity, 10^{-5}, of the training algorithm was reached. The MSE error after the training is 0.145352, and the total amount of time needed to train the network (initialization and training) was 1.53 seconds.

Classification Power of the Full and Reduced Models

In this section the predictive and classification power of the wavelet network are evaluated. More precisely, first the full model, including all nine attributes, is tested using the leave-one-out cross-validation. Then a comparison is made against the reduced model, which uses only six attributes.

The full model is first trained using all training examples. The classification matrix of the wavelet network in-sample is presented in Table 11.8. The wavelet network accuracy in the sample is 97.66%. The sensitivity is 97.07% and the specificity is 98.74%. Also, the wavelet network classified the malignant tumors incorrectly only three times, indicating a rate of failure of only 1.26%. Finally, in Table 11.9 we observe that the wavelet network classification ability is significantly greater than chance.

TABLE 11.8   In Sample Classification Matrix of the Full Model

Target	Forecast			Sensitivity	Specificity
	Benign	Malignant	Total		
Benign	431	13	444	97.07%	98.74%
Malignant	3	236	239	Rate of Missing Chances	Rate of Failure
Total	434	249	683	2.93%	1.26%

TABLE 11.9   Evaluation of the Classification Ability of the Full Wavelet Network

Maximum Chance	1.25% Max. Chance	Pro	Press's Q^a	Hit Ratio
65%	81.25%	54.50%	620.49	97.66%

[a]Press's Q critical values at confidence levels 0.1, 0.05, and 0.01: 2.71, 3.84, 6.63.

TABLE 11.10   Out-of-Sample Classification Matrix of the Full Model

Target	Forecast			Sensitivity	Specificity
	Benign	Malignant	Total		
Benign	431	13	444	97.07%	98.32%
Malignant	4	235	239	Rate of Missing Chances	Rate of Failure
Total	435	248	683	2.93%	1.68%

Next, the predictive power of the wavelet network is evaluated out-of-sample using the leave-one-out cross-validation method. Each time a validation sample is created that consists of only one observation, with the remaining pairs (**x**, *y*) used for the training of a wavelet network. In the next step, another validation sample is created and a new wavelet network is trained. The procedure is repeated until the wavelet network classifies all pairs (x, *y*). Table 11.10 presents the out-of-sample classification matrix of the full model. The accuracy of the full model out-of-sample is 97.51%, while the sensitivity and specificity are 97.07% and 98.32%, respectively. The misclassification of malignant cases is only 4, indicating a rate of failure of only 1.68%. Again, the very high hit ratio and the high Press's Q-statistic presented in Table 11.11 indicate that the wavelet network's classification ability is statistically significantly better than chance.

Next, the predictive power of the reduced model is evaluated in-sample and out-of-sample. The classification matrix is presented in Table 11.12. In-sample the sensitivity

TABLE 11.11   Evaluation of the Classification Ability of the Full Wavelet Network Out-of-Sample

Maximum Chance	1.25% Max. Chance	Pro	Press's Q^a	Hit Ratio
65%	81.25%	54.50%	616.29	97.51%

[a]Press's Q critical values at confidence levels 0.1, 0.05, and 0.01: 2.71, 3.84, 6.63.

TABLE 11.12   In-Sample Classification Matrix of the Reduced Model

	Forecast				
Target	Benign	Malignant	Total	Sensitivity	Specificity
Benign	431	13	444	97.07%	98.32%
Malignant	4	235	239	Rate of Missing Chances	Rate of Failure
Total	435	248	683	2.93%	1.68%

TABLE 11.13   Evaluation of the Classification Ability of the Reduced Wavelet Network

Maximum Chance	1.25% Max. Chance	Pro	Press's Q^a	Hit Ratio
65%	81.25%	54.50%	616.29	97.51%

[a]Press's Q critical values at confidence levels 0.1, 0.05, and 0.01: 2.71, 3.84, 6.63.

TABLE 11.14   Out-of-Sample Classification Matrix of the Full Model

	Forecast				
Target			Total	Sensitivity	Specificity
Benign	431	13	444	97.07%	97.91
Malignant	5	234	239	Rate of Missing Chances	Rate of Failure
Total	436	237	683	2.93%	2.09%

and specificity of the wavelet network are 97.07% and 98.32%, respectively. The malignant cases were misclassified only four times, indicating a very small rate of failure of 1.68%. Finally, in Table 11.13 we observe that Press's Q-statistic is higher than the critical values. Finally, the hit ratio is 97.51%, so the forecasting ability of the wavelet network is significantly better than chance.

Finally, the classification matrix of the reduced model out-of-sample is presented in Table 11.14. A closer inspection of the table reveals that the sensitivity and specificity are 97.07% and 97.91%. Also, there are five misclassified malignant cases, indicating a rate of failure of only 2.09%. Finally, examining Table 11.15, we conclude that the wavelet network has better forecasting ability than chance, with a hit ratio of 97.36%. Hence, the full model outperforms the reduced model by only one correct classification.

TABLE 11.15   Evaluation of the Classification Ability of the Reduced Wavelet Network Out-of-Sample

Maximum Chance	1.25% Max. Chance	Pro	Press's Q^a	Hit Ratio
65%	81.25%	54.50%	612.90	97.36%

[a]Press's Q critical values at confidence levels 0.1, 0.05, and 0.01: 2.71, 3.84, 6.63.

TABLE 11.16   Classification Power of the Full and Reduced Models[a]

Model	HU	Accuracy	Epochs	Correct	Wrong	B/B	M/B	M/M	B/M
Full (out)	1	97.51%	146	666	17	431	4	235	13
Full (in)	1	97.66%	146	667	16	431	3	236	13
Reduced (out)	3	97.36%	265	665	18	431	5	234	13
Reduced (in)	3	97.51%	264	666	17	431	4	235	13

[a]The algorithm concluded in four steps. In each step the following are presented: which variable is removed, the number of hidden units for the particular set of input variables and the parameters used in the wavelet network, the empirical loss and the prediction risk. (in), in-sample; (out), out-of-sample using leave-one-out cross-validation; B/B, case is B/WN predicts B; B/M, case is B/WN predicts M; M/M, case is M/WN predicts M; M/B, case is M/WN predicts B.

It is clear that the accuracy of the network remains practically the same even though three classifiers were removed from the data. Hence, we can conclude that the information that comes from the uniformity of cell shape, marginal adhesion, and mitoses does not contribute significantly toward classifying breast tumors, since the additional accuracy is only 0.15%. A summary of our results of the full and reduced models is presented in Table 11.16.

Our results indicate that a wavelet network can be used successfully in breast cancer classification, providing high classification accuracy. Moreover, the accuracy of the wavelet network is higher than those presented in relevant studies (Duch and Adamczak, 1998; Hassanien and Ali, 2006; Senapati et al., 2011; Setiono and Liu, 1997; Wei and Billings, 2007).

PART C: CLASSIFICATION OF BREAST CANCER (CONTINUED)

Two different methods were used in parts A and B to build a model for classifying breast cancer. In the first part, the data set were split into two samples, the training sample and the test sample, while in the second part, the cross-validation method was used to create additional sample to train and evaluate our model.

Another difference between the two methodologies was the corresponding input set of variables. In part A we assumed that all variables were statistically significant and used as predictors, while in part B we applied the variable selection algorithm to find which of the explanatory variables are statistically significant. Our results indicate that the uniformity of cell shape, marginal adhesion, and mitoses should be removed from the input of the training sample in breast cancer classification.

In this section we again split the data into training and validation samples, as in part A, but only the statistically significant variables will be used for construction of the wavelet networks. Hence, the attributes clump thickness, uniformity of cell size, single epithelial cell size, bare nuclei, bland chromatin, and normal nucleoli are used as input variables.

Classification

In-Sample   The classification matrix of the training sample is presented in Table 11.17. Close inspection of the table reveals very good classification rates. More specifically, the wavelet network classified correctly 274 benign cases and 182

TABLE 11.17   In-Sample Classification Matrix

	Forecast				
Target	Benign	Malignant	Total	Sensitivity	Specificity
Benign	274	10	284	96.48%	93.81%
Malignant	12	182	194	Rate of Missing Chances	Rate of Failure
Total	286	182	478	3.52%	6.19%

malignant cases. Hence, the wavelet network classified correctly 456 of 478 cases (95.40%). The specificity of the models is 93.81% and the sensitivity is 96.48%. On the other hand, the rate of failure and the rate of missing chances are very low, 6.19% and 3.52%, respectively. Finally, the fitness function is 0.5503. Comparing our results to those in part A, we observe that the wavelet network with the reduced set of input variables misclassified 12 malignant cases, whereas when the full set of input variables was used, the wavelet network misclassified only 7 cases.

Evaluation of the classification ability of the wavelet network is presented in Table 11.18. The maximum chance criterion is 59.41%, whereas in the heuristic method presented by Hair et al. (2010) it is 74.27%. Finally, the proportional chance criterion is 51.77%. The hit ratio, 90.40%, is significantly larger than the various chance criteria. Hence, the model is predicting significantly better than chance. This is confirmed by the large value of Press's Q-statistic, which is greater than the critical values at the 0.1, 0.05, and 0.01 confidence levels.

Out-of-Sample   Next, the forecasting and classification ability of the trained wavelet network are evaluated in the validation sample. The data of the validation sample were not used during the training phase. Hence, these are new data that were never presented to the wavelet network.

In the validation sample there are 160 benign cases given the value −1 and 45 malignant cases given the value 1. The classification matrix of the training sample is presented in Table 11.19. Close inspection of the table reveals perfect prediction ability and classification rates. More specifically, the wavelet network classified correctly 160 benign cases and 45 malignant cases. Hence, the wavelet network classified all 205 cases correctly (100%). Hence, both the sensitivity and the specificity of the model are 100%, whereas the rate of failure and the rate of missing chances are 0%. Finally, the fitness function, 0.6, is the maximum value possible. Comparing our results to those of part A, we conclude that although a poorer fit was obtained to the data in-sample when the truncated set of input variables was used, the predictive

TABLE 11.18   Evaluation of the Classification Ability of the Wavelet Network

Maximum Chance	1.25% Max. Chance	Pro	Press's Q^a	Hit Ratio
59.41%	74.27%	51.77%	394.05	95.40%

[a]Press's Q critical values at confidence levels 0.1, 0.05, and 0.01: 2.71, 3.84, 6.63.

TABLE 11.19 Out-of-Sample Classification Matrix.

Target	Forecast Benign	Malignant	Total	Sensitivity	Specificity
Benign	160	0	160	100%	100%
Malignant	0	45	45	Rate of Missing Chances	Rate of Failure
Total	160	45	205	0.00%	0.00%

TABLE 11.20 Out-of-Sample Evaluation of the Classification Ability of the Wavelet Network

Maximum Chance	1.25% Max. Chance	Pro	Press's Q^a	Hit Ratio
78.05%	97.56%	65.73%	205.	100%

[a]Press's Q critical values at confidence levels 0.1, 0.05, and 0.01: 2.71, 3.84, 6.63.

power of the wavelet network increased. The out-of-sample results provided a perfect classification.

Evaluation of the classification ability of the wavelet network is presented in Table 11.20. The maximum chance criterion is 78.05%, whereas in the heuristic method presented by Hair et al. (2010) it is 97.56%. Finally, the proportional chance criterion is 65.73%. The hit ratio, 100%, is significantly larger than the maximum chance and the proportional chance criteria. Hence, the model is predicting significantly better than chance. This is confirmed by the large value of Press's Q statistic, which is greater than the critical values at the 0.1, 0.05, and 0.01 confidence levels.

CONCLUSIONS

Mammography is probably the most effective method for breast tumor detection. In this chapter a computer-aided system for breast cancer classification was proposed. More precisely, in this chapter a nonlinear nonparametric wavelet neural network was constructed and trained to identify and classify benign and malignant breast cancer cases correctly. The data set were obtained by the UCI Machine Learning Repository and corresponds to clinical cases of breast cancer in Wisconsin. The classification was based on nine attributes: clump thickness, uniformity of cell size, uniformity of cell shape, marginal adhesion, single epithelial cell size, bare nuclei, bland chromatin, normal nucleoli, and mitoses.

Two modeling approaches were presented. In the first case the data were split into training and validation samples. The first set was used for the model selection training of the wavelet network; the second set was used to assess its classification power. The procedure was applied in the full set of explanatory variables and on the truncated set after the variable selection algorithm was employed. Our results indicate that the wavelet network can classify the clinical case with very high accuracy. When the full set of explanatory variables was used as an input, only one case was misclassified.

On the other hand, on the reduced input of variables the wavelet network was able to classify all cases perfectly.

In the second case, the hold-one-out cross-validation was used to create additional training and validation samples. At the same time, the model identification algorithm was used. Our results indicate that only six input variables—clump thickness, uniformity of cell size, single epithelial cell size, bare nuclei, bland chromatin, and normal nucleoli—can offer the same level of classification accuracy as the full set of input variables.

In every case the classification ability of the wavelet network was very high. The wavelet network had high generalization ability and produced robust and reliable results both in-sample and out-of-sample, indicating that wavelet networks can be accurate nonlinear nonparametric estimators for breast cancer recognition.

REFERENCES

Cheng, H. D., Shan, J., Ju, W., Guo, Y., and Zhang, L. (2010). "Automated breast cancer detection and classification using ultrasound images: a survey." *Pattern Recognition*, 43(1), 299–317.

Duch, W., and Adamczak, R. (1998). "Statistical methods for construction of neural networks." *Proceedings of 5th International Conference on Neural Inf*, 639–642.

El-Sebakhy, E. A., Faisal, K. A., Helmy, T., Azzedin, F., and Al-Suhaim, A. (2006). "Evaluation of breast cancer tumor classification with unconstrained functional networks classifier." *IEEE International Conference on Computer Systems and Applications*, 281–287.

Hair, J., Black, B., Babin, B., and Anderson, R. (2010). *Multivariate Data Analysis*, 7th ed. Prentice Hall, Upper Saddle River, NJ.

Hassanien, A., and Ali, J. (2006). "Rough set approach for classification of breast cancer mammogram images. In *Fuzzy Logic and Applications*, V. Di Gesú, F. Masulli, and A. Petrosino, eds. Springer-Verlag, Berlin, 224–231.

Mangasarian, O. L., and Wolberg, W. H. (1990). "Cancer diagnosis via linear programming." *SIAM News*, 23(5), 1–18.

Senapati, M., Mohanty, A., Dash, S., and Dash, P. (2011). "Local linear wavelet neural network for breast cancer recognition." *Neural Computing and Applications*, 1–7.

Setiono, R., and Liu, H. (1997). "NeuroLinear: from neural networks to oblique decision rules." *Neurocomputing*, 17(1), 1–24.

Wei, H.-L., and Billings, S. A. (2007). "Feature subset selection and ranking for data dimensionality reduction." *IEEE Transactions on Pattern Analysis and Machine Intelligence*, 29(1), 162–166.

Zainuddin, Z., and Ong, P. (2010). "Improved wavelet neural network for early cancer diagnosis using clustering algorithms." *International Journal of Information and Mathematical Sciences*, 6(1), 30–36.

Index

Wavelet Neural Networks: With Applications in Financial Engineering, Chaos, and Classification,
First Edition. Antonios K. Alexandridis and Achilleas D. Zapranis.
© 2014 John Wiley & Sons, Inc. Published 2014 by John Wiley & Sons, Inc.